My Lighting Handbook

My Lighting Handbook

Volume 1: Interior

Lorenzo Simoni

www.lorenzo-simoni.com

Printed by CreateSpace – an Amazon.com company
Available from Amazon, CreateSpace, my own website www.lorenzo-simoni.com and other retail and online stores

For Shanti and Walter

Contents

Preface to the third edition — xvii

Acknowledgements — xix

Part 1 — 3

1 What is light? — 5
- 1.1 Light ... 5
- 1.2 Infrared radiation ... 6
- 1.3 Ultraviolet radiation ... 8
- 1.4 Spectral power distribution ... 9
- 1.5 Self evaluation ... 10

2 Colour, colour temperature and colour rendering — 13
- 2.1 Spectral and perception colours ... 13
- 2.2 The observer ... 14
- 2.3 Elements of colour synthesis ... 17
- 2.4 Light and colour ... 18
- 2.5 Colour rendering ... 22
- 2.6 Colour temperature ... 24
    - 2.6.1 Reciprocal mega Kelvin ... 26
- 2.7 Self evaluation ... 30

3 Basic electrical quantities — 33
- 3.1 Electrical current ... 33
- 3.2 Difference of electrical potential, voltage ... 33
- 3.3 Electrical power ... 34
- 3.4 Self evaluation ... 35

4 Luminous flux   37

    4.1 Definition of luminous flux. 37

    4.2 Luminous efficacy 38

    4.3 Light output ratio (R_{LO}). 42

    4.4 Total efficacy 43

    4.5 Luminaire's luminous efficacy (LLE) 44

    4.6 Self evaluation 45

5 Luminous intensity   47

    5.1 Photometric solid 50

    5.2 C, gamma system of reference 51

    5.3 Light intensity distribution curve. 53

    5.4 Beam spread 56

    5.5 Self evaluation 59

6 Illuminance and evaluation of a lighting system   61

    6.1 Visual task 61

    6.2 Task area 62

    6.3 Concept of illuminance. 62

    6.4 Vertical and horizontal illuminance 64

    6.5 Cylindrical illuminance. 64

    6.6 Semi-cylindrical illuminance. 65

    6.7 Evaluation of a lighting system: performance 65

    6.8 Maintenance factor 66

    6.9 Average maintained cylindrical illuminance ($\bar{E}_{Zm}$) 67

    6.10 Illuminance uniformity (U_0). 68

    6.11 Modelling 70

    6.12 Isolux curves 73

        6.12.1 Position of a particular level of illuminance 74

        6.12.2 Rate of change of illuminance on the surface. 75

6.12.3 Range of values for the illuminance in a certain spot.	76
6.13 Self evaluation	76

7 Intensity and illuminance — 79

7.1 Illuminance under a luminaire as a function of distance and luminous intensity	79
7.2 Illuminance and intensity: general formula	82
7.3 Cone diagrams	84
7.4 Illuminance uniformity and light intensity distribution	86
7.5 Self evaluation	88

8 Elements of reflection — 91

8.1 Reflectance	91
8.2 Types of reflection	92
8.2.1 Specular reflection	92
8.2.2 Diffuse reflection	93
8.2.3 Mixed reflection	94
8.3 Reflection diagram	94
8.4 Self evaluation	95

9 Luminance — 97

9.1 Luminance	97
9.2 Luminance contrast	100
9.3 Self evaluation	102

10 Visual Comfort — 105

10.1 Visual comfort	105
10.2 Veil	105
10.3 Shade	107
10.4 Flicker	108
10.5 Glare	109
10.5.1 Disability glare	109
10.5.2 Discomfort glare	110

10.6 Bright lamps. 111

10.7 Luminance control for workstations with Display Screen Equipment 112

10.8 Offending zone 115

10.9 Unified Glare Rating (UGR) 116

10.10 Self evaluation 119

Part 2   121

11 How is light produced and evaluated   123

11.1 Light production. 123

    11.1.1 Incandescence 123

    11.1.2 Fluorescence. 123

    11.1.3 Electric arc 124

    11.1.4 Semiconductor physics 124

    11.1.5 Types of lamps and processes 125

11.2 Evaluation of a lamp 126

    11.2.1 The shape 126

       11.2.1.1 The bulb 126

       11.2.1.2 The cap or base 126

    11.2.2 Correlated Colour Temperature 127

    11.2.3 Colour rendering 127

    11.2.4 Luminous Efficacy 128

    11.2.5 Average rated life (B_{50}) 128

    11.2.6 Electrical devices 131

11.3 Self evaluation. 132

12 Halogen lamps   135

12.1 Notes on incandescent lamps 135

12.2 Types of lamps 135

12.2.1 Lamps without reflector	136
12.2.2 Lamps with reflector	136
12.2.2.1 Halogens with dichroic reflector	137
12.3 Description	137
12.3.1 Inner workings and halogen cycle	138
12.4 Evaluating grid	139
12.4.1 Bulb and cap	139
12.4.2 Colour temperature	139
12.4.3 Colour rendering	139
12.4.4 Luminous efficacy	140
12.4.5 Average rated life	140
12.4.6 Electrical devices	140
12.5 Considerations	140
12.5.1 Strengths	140
12.5.2 Weaknesses	141
12.6 Examples of fixtures	141
12.7 Self evaluation	142

13 Fluorescent lamp   145

13.1 Lamp types	145
13.2 Description	145
13.2.1 Inner workings	147
13.3 Evaluating grid	147
13.3.1 Bulb and cap	147
13.3.2 Colour temperature	147
13.3.3 Colour rendering	148
13.3.4 Luminous efficacy	148
13.3.5 Average rated life	148
13.3.6 Electrical devices	148

13.4 Traditional Fluorescent lamps. 149
    13.4.1 Considerations. 149
    13.4.2 Example of fittings 150
13.5 Compact integrated fluorescent lamps 150
    13.5.1 Cost comparison to the halogen lamp. 150
13.6 Compact non-integrated fluorescent lamps 153
    13.6.1 Considerations. 153
    13.6.2 Example of light fittings. 153
13.7 Self evaluation 154

14 Metal halide lamps 157

14.1 Types of lamps 157
14.2 Inner workings 157
14.3 Totally enclosed luminaires only 158
14.4 Start up and warm restart time 159
14.5 Cyclic behaviour at end of life 159
14.6 Cost comparison to the halogen lamp. 159
14.7 Evaluating grid 162
    14.7.1 Bulb and cap 162
    14.7.2 Colour temperature 162
    14.7.3 Colour rendering index 162
    14.7.4 Luminous efficacy. 162
    14.7.5 Average rated life 163
    14.7.6 Electrical devices 163
14.8 Considerations. 163
    14.8.1 Strengths 163
    14.8.2 Weaknesses 164
14.9 Example of light fittings. 164
14.10 Self evaluation 166

15 LEDs — 169

- 15.1 LED lamps . . . 169
  - 15.1.1 Comparison to halogens. . . . 170
- 15.2 White light LED modules. . . . 172
- 15.3 Coloured light LED modules . . . 172
- 15.4 Description . . . 172
  - 15.4.1 Inner workings . . . 173
  - 15.4.2 Heat dissipation . . . 174
  - 15.4.3 Stable driver. . . . 174
- 15.5 Average life . . . 175
- 15.6 Binning . . . 175
- 15.7 Luminance control . . . 177
- 15.8 Evaluating grid . . . 177
  - 15.8.1 Bulb and cap . . . 177
  - 15.8.2 Colour temperature . . . 178
  - 15.8.3 Colour rendering . . . 178
  - 15.8.4 Luminous efficacy. . . . 178
  - 15.8.5 Average life . . . 179
  - 15.8.6 Electrical devices . . . 179
- 15.9 OLEDs . . . 179
- 15.10 Example of fittings. . . . 180
- 15.11 Self evaluation . . . 180

16 Economic comparison between lamps — 183

Part 3 — 185

17 Luminaires: general description — 187

- 17.1 Functions and parts of a luminaire . . . 187

	17.2	Ingress protection and the IP rating	189
	17.3	Visual task and activities	191
	17.4	Self evaluation	195

18 Selecting by flux: CIE classification   197

     18.1 Sample room 197
     18.2 Direct 198
     18.3 Indirect 200
     18.4 Semi-direct 203
     18.5 Semi-indirect 204
     18.6 General-diffuse 206
     18.7 Direct-indirect 208
     18.8 Self evaluation 210

19 Selecting by LID: Axially symmetric   213

     19.1 Sample room 213
     19.2 Axially symmetric luminaires 213
     19.3 Axially symmetric narrow beam 215
         19.3.1 Application: single fitting in sample interior 216
         19.3.2 Application: sample interior with realistic lighting 217
         19.3.3 Application: horizontal illuminance 218
         19.3.4 Application: vertical illuminance 219
         19.3.5 Application: semi-cylindrical illuminance 223
         19.3.6 Application: cylindrical illuminance 224
         19.3.7 Application: Spherical illuminance 225
     19.4 Axially symmetric wide beam 226
         19.4.1 Application: single luminaire in the sample interior 226
         19.4.2 Application: sample interior with realistic lighting 228
     19.5 How to dimension an axially symmetric luminaire 229
     19.6 Self evaluation 231

20 Selecting by LID: symmetric about two planes   **233**

    20.1 Symmetric about two planes: narrow beam. 233

    20.2 Symmetric about two planes: wide beam. 234

        20.2.1 Application: single fitting in the sample interior 234

        20.2.2 Application: Sample interior with realistic lighting. 237

        20.2.3 How to select the right orientation 238

            20.2.3.1 Luminance control 240

    20.3 Self evaluation 241

21 Selecting by LID: asymmetric   **243**

    21.1 Traditional asymmetric 244

        21.1.1 Application: single fitting to sample interior 246

        21.1.2 Application: Sample interior with realistic lighting. 248

    21.2 Wall washers 248

        21.2.1 Application: narrow beam wall-washer and sample interior 250

        21.2.2 Application: wide beam wall-washer and the sample interior 252

        21.2.3 Application: Sample interior with realistic lighting. 254

    21.3 Fully asymmetric luminaire. 255

    21.4 Self evaluation 256

22 Selecting the fitting by other criteria   **259**

    22.1 Selecting by lamp 259

        22.1.1 Colour rendering 259

        22.1.2 Colour temperature 260

        22.1.3 Size requirements 260

        22.1.4 Instant start 261

        22.1.5 Dimming 261

        22.1.6 Coloured light 261

    22.2 Selecting by luminance control 261

    22.3 Selecting by efficacy - Specific Connected Load 262

22.4 Selecting a luminaire by beam spread 264
22.5 Selecting a luminaire by mounting 266
    22.5.1 Ceiling mounting 266
    22.5.2 Wall mounting. 267
    22.5.3 Suspended mounting 268
    22.5.4 Portable mounting and free standing 269
    22.5.5 Recessed mounting 270
    22.5.6 Track mounting 270
22.6 Selecting a light fitting by commercial classification 271
    22.6.1 Downlight 271
    22.6.2 Troffer. 272
    22.6.3 Commercial fluorescent. 273
        22.6.3.1 Wraparound diffuser 273
        22.6.3.2 Fluorescent strip - batten 274
    22.6.4 Industrial fixtures 275
    22.6.5 Linear systems 276
    22.6.6 Wall-washers (commercial definition) 276
    22.6.7 Wall grazer 277
    22.6.8 Accent fixtures 278
    22.6.9 Cove lights 279
    22.6.10 Decorative lights. 280
    22.6.11 Task lights 280
22.7 Self evaluation 281

Part 4   283

23 The concept   285

23.1 What is there to see? 286
    23.1.1 Attributes 287

23.1.2 Example of use 288
23.1.3 Luminance and surfaces. 289
23.1.4 Photo realistic rendering 291
23.2 From where should the light come?. 292
23.3 Sharpness and flow 292
23.3.1 Single point-like source 294
23.3.2 Multiple point-like sources 294
23.3.3 Diffused light from one side 294
23.3.4 Fully diffused light 295
23.4 What kind of light should we use? 295
23.5 Self evaluation 295

24 The interior   299

24.1 Architectural features 299
24.2 Materials and finishing of the surfaces 300
24.2.1 Protection 300
24.2.2 Luminance. 300
24.2.3 Reflectance 301
24.2.4 Colour bleeding 301
24.3 Movement of the users in the interior. 301
24.4 Self evaluation 302

25 The activities   305

25.1 Interview with the client 305
25.2 The activity table 305
25.3 Identification of the task areas 306
25.4 Compliance 307
25.5 Self evaluation 307

26 Layered design   309

26.1 Decorative layer. 309

26.2 Task layer 310

26.3 Brief note on regulations and standards. 311

26.4 General layer 312

26.5 Self evaluation 312

Part 5   315

27 Introduction to a lighting calculation program: DIALux   317

27.1 The objective 317

27.2 The programme 318

     27.2.1 3D modeller 318

     27.2.2 Database 321

     27.2.3 Simulator 322

     27.2.4 Output package 322

27.3 Self evaluation 324

28 DIALux Wizards   327

28.1 Use of a wizard 327

28.2 An example: the supermarket 329

28.3 Self evaluation 330

29 A sample interior   333

29.1 Let's build the interior. 333

29.2 Objective 333

29.3 Create the room geometry 334

29.4 Add doors and window 334

29.5 Textures 335

29.6 Insert furniture 335

29.7 Surroundings of the building 337

30 Calculation surfaces   341

30.1 Walls, floor and ceiling 341
30.2 Work plane 341
30.3 Surface of an object 342
30.4 Free surfaces 343
30.5 Task area and surrounding area 343
30.6 Self evaluation 344

31 Control groups and light scenes   347

31.1 What is a light scene? 347
31.2 What is a control group? 348
31.3 An example: lighting a restaurant 349
31.4 Application to our sample interior 352
31.5 Self evaluation 352

32 Daylighting   355

32.1 Daylighting in DIALux 355
32.2 Daylight factor 356
32.3 Calculation 357
32.4 Evaluation of the results. 358
32.5 False colour display 358
32.6 Daylight factor isolines 360
32.7 Daylight screening 363
32.8 Self evaluation 364

33 Luminaires in DIALux   367

33.1 Insertion into our sample project 367
33.2 Positioning 369
33.3 Definition of the necessary scenes and control groups 371
33.4 Self evaluation 374

34 Calculation, evaluation and improvement   377

34.1 Evaluation of the results. 377

	34.1.1 Daylight and artificial 377
	34.1.2 Artificial 381

- 34.2 Possible improvements 382
- 34.3 Work flow for modifying a system 384
  - 34.3.1 Output the results for the surface 387
  - 34.3.2 Evaluate the amount of illuminance we need to add 387
  - 34.3.3 Duplicate the room, delete luminaires, place new luminaires and calculate the result. 388
  - 34.3.4 Final result 388
- 34.4 UGR evaluation 389
- 34.5 Self evaluation 392

Alphabetical list of topics   **395**

35 Bibliography   **403**

Preface to the third edition

This time my effort has been to better organise chapter in more coherent units, to try an be clearer on certain points and to remove repetitions and technicalities in the text.

I also converted the whole book to grey-scale, to reduce the printing costs, and this was a challenge, because the book contains about 250 images, many of which required a special colour to grey balancing to maintain their usefulness and meaning (e.g. coloured object lit by coloured light).

I can only hope the result is effective.

Acknowledgements

The second edition was perpetrated by the same culprit of the first one.

I would like to gratefully acknowledge the kindness of Philips Italia, as they allowed me to use the images of their products, and that helped make this book clearer and more effective.

I would also like to thank DIAL GmbH for allowing me to use any images I wanted from their lighting calculation software, DIALux, which I taught in part 5 of the book and with which all the calculations are made.

The graphics, diagrams, models, photographs, pagination, editing, proofing, cover and of course all the mistakes are made entirely by me.

Lorenzo Simoni
Milano, 2015

Part 1

This section of the book is dedicated to the fundamentals of lighting science, some very basic informations about colour theory, lighting standards and electrical engineering, and a more detailed look into visual comfort.

What is light?

1

A few details on what light is can help a designer in their projects.

1.1 Light

Light is one among many different types of electromagnetic radiations. An electromagnetic radiation is what happens when an electric or a magnetic field changes in time or in space.

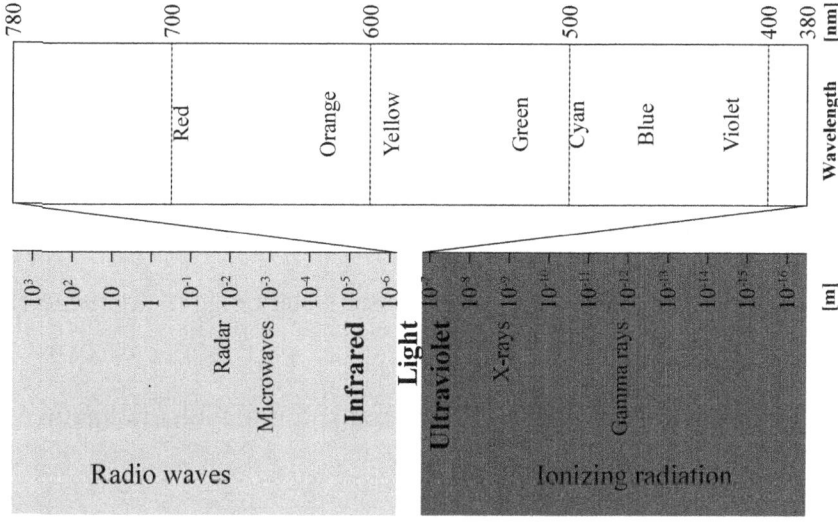

Figure 1: Electromagnetic spectrum.

There are many different types of electromagnetic radiations, each defined by a certain frequency. In order of increasing frequency, we find: radio waves, (microwaves and radar transmission belongs to this category), infrared radiation, light, ultraviolet radiation, x-rays and gamma rays (Figure 1).

It is important to know that frequency and wavelength are inversely proportional to one another: as the wavelength of a radiation increases, its frequency will decrease.

The energy content of a radiation is an indicator of how damaging the exposure to that radiation will be, and it is proportional to its frequency. In layman's terms, this means that we can listen to the radio almost anywhere (i.e. we are constantly immersed in radio waves) without any risk, because radio waves have very little energy (non ionizing radiations); however when we need x-rays taken, we must use precautions, because x-rays are very energetic and can damage our body (ionizing radiations).

We define light as follows:

> `Light is the part of the electromagnetic spectrum that can be seen by the average observer.` (1)

Its wavelength goes from about 380 nm[1] to about 780 nm. Light is only a small part of the total radiation emitted, the vast majority of the electromagnetic spectrum is invisible.

Because light is what each observer sees, the range of frequencies changes (slightly) with each person.

In lighting we are interested in infrared and ultraviolet radiation as well.

1.2 Infrared radiation

Infrared (IR) is the invisible radiation whose frequency is shorter (infra) than the frequency of red light. This means that the wavelength is longer: from about 700 nm to about 1 mm (Figure 1).

But what is infrared radiation, then?

---

1. A nanometre is a billionth of a meter (10^{-9} meter). For reference, the thickness of a human hair is a hundred thousand times larger than a nanometre.

A simplified definition, that works perfectly well for all our intents and purposes follows.

> `Infrared radiation` is heat. (2)

Heat can be transferred in three ways:

1. By conduction, that is by direct contact between a warmer object and a cooler one, such as when you burn your finger touching a hot pot filled with boiling water and pasta.
2. By convection i.e. by heating a fluid whose motion will transfer it to another object; such as when boiling water transfers the heat of the hob to the pasta in the pot and cooks it.
3. By radiation, i.e. a warm object will emit radiation and cool itself, such as when you are heated by sunlight as you are sitting on a sun chair after eating too much pasta.

Figure 2: Philips PAR38 IR: a lamp that emits infrared radiations for industrial applications.

IR emitted by a lamp are a waste for two reasons: they are invisible, so all the electrical energy used to produce them is not helping people see; and they increase the amount of heat that the building's air conditioning system must eliminate.

Some tasks are sensitive to heat or particularly delicate, and might be ruined by the full IR emission of a halogen lamp. In 15.1.3 we shall see what can be done to deal with this problem.

Other applications require IR: the hot-food display cabinet in a supermarket keeps food warm by using lamps such as the one in Figure 2.

1.3 Ultraviolet radiation

Ultraviolet (UV) is the radiation whose frequency is larger (ultra) than that of violet light. This of course implies that the wavelength is shorter (from about 10 nm to 380 nm).

Figure 3: UV degradation in dyed textiles exposed to sunlight.

Exposure to UV radiation has few benefits, usually only available in medical or strictly controlled conditions, and many risks.

Among the benefits for human beings there is the increase in production of vitamin D, the treatment of skin conditions like psoriasis, and the aesthetic tanning of the skin.

The risks[1] are much more relevant. UV radiation can damage our DNA, which causes skin cancer (melanoma, basal cell carcinoma or squamous cell carcinoma), premature aging and other skin conditions, immune system suppression, and eye damage (cataracts, pterygium, skin cancer around the eyes and degeneration of the macula).

UV radiations also have detrimental effects on inanimate objects: UV degradation is the result of exposing plastic polymers and dyes to ultraviolet radiation. Plastic becomes brittle and fragile, while pigments and dyes lose saturation (Figure 3).

Colour degradation is extremely problematic in museums, because precious objects are on display. Paintings and textiles in particular are quite susceptible to this kind of damage. This is the reason why UV filters are compulsory in these installations, for both luminaires and windows.

1.4 Spectral power distribution

Light works both as a particle and as a wave but, as lighting designers, we deal mostly with the wave, because it explains many properties of high quality lighting: colour rendering (2.5 on page 22), colour temperature (2.6 on page 24), and the colour of light (on page 17).

In a manner similar to how a chord played on a guitar contains the vibration of multiple strings at different frequencies, white light contains all the colours of the rainbow in different relative quantities (additive synthesis).

Lamp manufacturers publish the spectral power distribution of their lamps (Figure 4). It shows how much power per surface unit is emitted at each frequency or, in simpler terms, how much of each colour is present in the light.

---

1. See e.g. UNITED STATES ENVIRONMENTAL PROTECTION AGENCY 2010

In our example, it would mean how much of each note produced by a certain vibrating string is part of the chord.

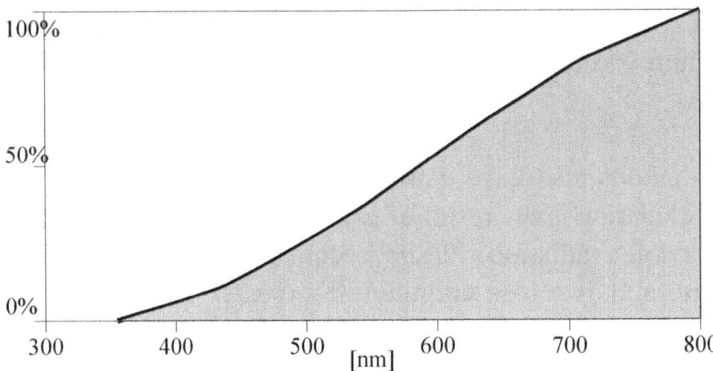

Figure 4: Spectral power emission of an incandescent lamp.

The spectral power distribution of a lamp shows the characteristics of the light: looking at Figure 4, it is clear that this kind of lamp will not emit a lot of power in the blue region of the spectrum (around 380 nm of wavelength), compared to the red region (around 780 nm). This means that the light emitted will contain a lot of red and not very much blue, and it will make red objects shine out more than blue ones, changing the hue of the light towards yellow.

1.5 Self evaluation

Answer the following questions to test your understanding of the material:

1. Give a complete definition of light.
2. Explain what is Infrared radiation and what it does.
3. Explain what is UV radiation and what it does.
4. Explain what is the spectral power distribution diagram and what it is for.
5. Can you tell what colour will the light be from the spectral power distribution?

Colour, colour temperature and colour rendering

2

What is colour

The colour of an object is the result of the interaction between:

1. The light that illuminates the object.
2. The surface of the object.
3. The eye of the observer.

Let us spend a little time looking into points 1 and 3.

2.1 Spectral and perception colours

In general we can divide colours into two categories: spectral colours and perception colours.

Spectral colours correspond to a precise wavelength of monochromatic light, so that we can surely say that light at 610 nm of wavelength will look orange to the eye (Table 1 on page 16). The spectral colours are displayed in the curved border of the CIE[1] xy chromaticity diagram (Figure 5).

Perception colours are obtained mixing couples of spectral colours. For example (Figure 5), the colour in O can be obtained by mixing (in appropriate amounts) the colours A and B, or C and D, or E and F. All the perception colours are displayed in the inner part and bottom straight border of the chromaticity diagram.

---

1. CIE (Commission international de l'Eclairage) is a worldwide organisation devoted to the advancement of knowledge and the production of standards to improve the lighted environment.

The concept of perception colours is applied to computer monitors and coloured lighting: both can produce millions of colours using only three[2] types of emitters: a red one, a green one and a blue one.

Figure 5: CIE chromaticity diagram.
Mixing appropriate quantities of colour A and colour B, or colour C and colour D, or colour E and colour F produces the same result: colour O.

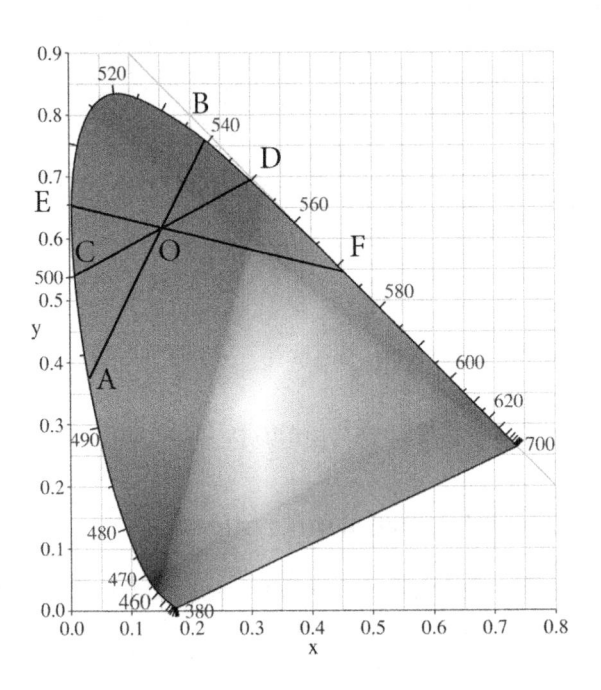

2.2 The observer

Radiometry is the study and measurement of electromagnetic radiation, and it deals mostly with the distribution of its power in space, but it does not concern itself with the observer. On the other hand photometry and illuminating science are founded on the interaction of the radiation produced with the eye of an average person.

This average observer has been defined by the CIE since 1924 (CIE 1924; SCHREUDER 2008), and it is represented by a curve called CIE photopic spectral luminous efficiency function (see Figure 7). It shows how sensitive the average eye is to the different wavelengths of light.

---

2. Sometimes there are four emitters, the one added producing a non primary colour, e.g. cyan.

The curve has a maximum at 555nm of wavelength, equivalent to yellowish green, and decreases quickly from there.

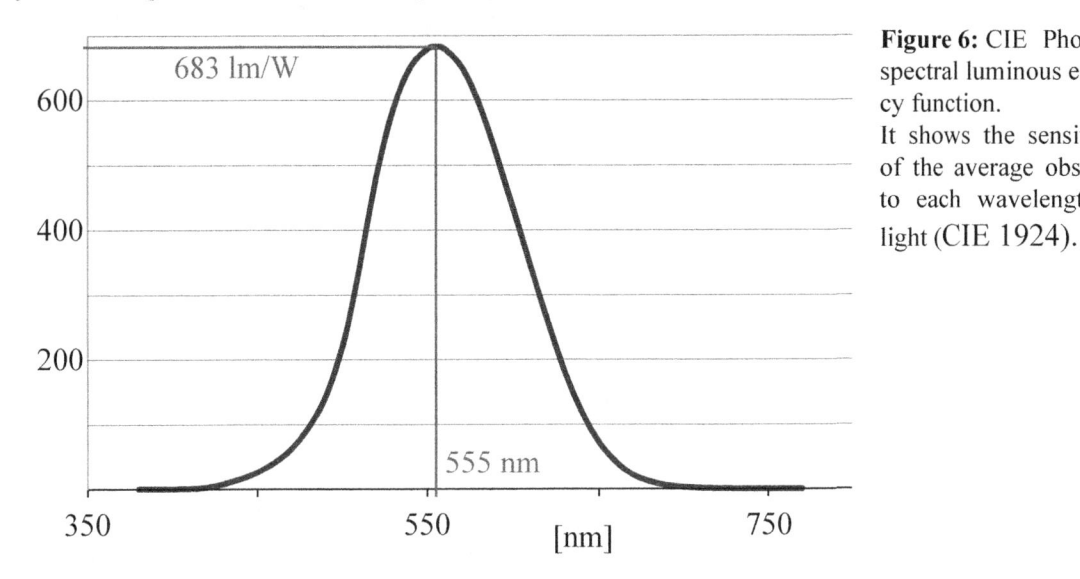

Figure 6: CIE Photopic spectral luminous efficacy function.
It shows the sensitivity of the average observer to each wavelength of light (CIE 1924).

Remembering that each wavelength is perceived as a pure hue of a primary colour (spectral colour), and using the major spectral colour definitions, we can express the curve as a table (Table 1) that correlates the colour with the sensitivity of the average human eye to that colour.

Looking at the table, we can see that a red light emitting a unit of power will appear to us about a hundred times dimmer than an equally powerful green light. This is because the green light produces a 0.86 effect in the eye, while the red light produces a 0.0082 effect, which is about a hundred times smaller.

Using energy efficiently is very important, and with Table 1 we can already answer an important question: what is the most efficient colour, i.e. the colour that produces the maximum sensation of brightness using the minimum amount of energy. The answer is the colour at which the eye sensitivity is maximum, i.e. monochromatic light at 555 nm, sort of a yellowish green colour.

Table 1: Sensitivity of the average observer to the different wavelengths of light (CIE 1924).

Colour	Wavelength	Sensitivity (max = 1)
Red	690	0.0082
Orange	610	0.50
Yellow	580	0.87
Green	530	0.86
Blue	470	0.091
Indigo	430	0.012
Violet	410	0.0012

Why then isn't everything illuminated with this light? The answer (colour rendering) is in 2.5 on page 22.

Finally I need to talk about the "photopic" adjective added to the diagram and table. Because of the physiology of the human eye, our vision mainly works in two "modes": a low light mode, called scotopic, and a high light mode, called photopic.

> **Photopic vision** is day-time vision (SCHREUD-ER 2008), with the observer fully adapted to high levels of luminance (i.e. brightness). There is full colour perception. (3)

This mode of vision has the highest visual acuity, it comes from cones (a type of sensor cell in the retina), and it is normally used for lighting design.

> **Scotopic vision** is night-time vision, with the observer fully adapted to low levels of luminance. There is no colour perception. (4)

Colour, colour temperature and colour rendering

This mode comes from rods (the other type of light perceptor in the retina), with much lower visual acuity. Consequently there is a different sensitivity curve[3], but it is not used in lighting design.

Between the two, there is a third mode, explored much later and still considered sort of a work in progress[4], called mesopic vision. This happens when both cones and rods are working at the same time. It can be useful for lighting, in particular in low light outdoor areas, but it is beyond the scope of the book.

2.3 Elements of colour synthesis

We recognize two types of colour synthesis: additive and subtractive (see Figure 7).

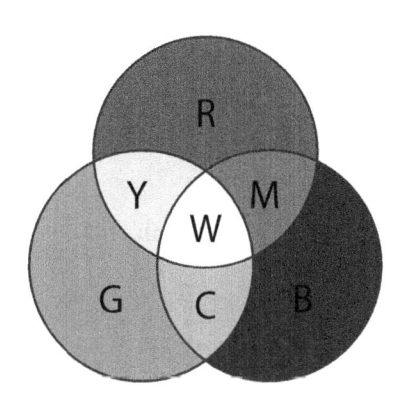

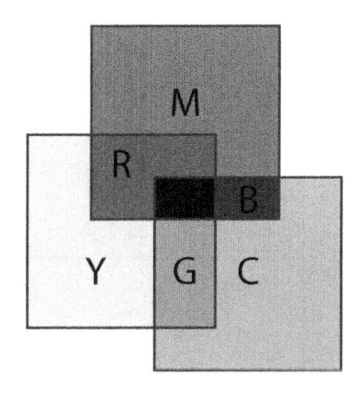

Figure 7: Additive colour synthesis (left) and subtractive colour synthesis (right).
R is for red, G for green, B for blue, M for magenta, C for cyan and Y for yellow.

Additive colour synthesis, in simple terms, describes how coloured light behaves.

If, for example, we shine three partially overlapping beams of light on a white screen: a red, a green and a blue one, the result will be similar to the left part of Figure 7. The area where red and green overlap will look yellow, the one where red and blue overlap will be magenta, and the one where blue and green overlap will produce

---

3. CIE 1951

4. CIE 1989

cyan. The area in the centre where all three colour overlap will be white.

Subtractive colour synthesis, on the other hand, portrays the behaviour of coloured filters.

Table 2: Additive colour synthesis table.

Source	Source	Result
Red	Green	Yellow
Red	Blue	Magenta
Green	Blue	Cyan

If we have a magenta light source, and we put a yellow filter over it, we obtain red light. The same result comes from a yellow light source and a magenta filter. If we replace the magenta filter with a cyan one, we get blue. If we subtract from any source (let's say magenta) both the others (yellow and cyan), we filter all the light and get darkness.

Table 3: Subtractive colour synthesis table.

Source	Filter	Result
Magenta	Yellow	Red
Magenta	Cyan	Blue
Yellow	Cyan	Green

2.4 Light and colour

When the light changes, the same object will look different in a small or big way according to the amount of variation. Let's show this by illuminating a still life with different light sources.

First we illuminate the dish with a candle (Figure 8): the warm and yellowish light highlights the yellow and red fruits, and all the colours are visible. The atmosphere is cosy and relaxed.

Colour, colour temperature and colour rendering 19

We then use a fluorescent cool light (Figure 9): the colours change slightly, but they are all there. The atmosphere is now more stimulating.

Figure 8: Still life lit with a candle. Notice the warm tones and the highlighting of the tomato.

Figure 9: Still life lit with a fluorescent. Notice the different atmosphere.

When we illuminate the dish with coloured light, things change dramatically, according to colour theory.

Figure 10: Still life lit by blue light.
Notice how the red tomato appears almost black, because it contains no blue.

Blue light

Figure 11: Still life lit by red light.
The tomato looks about the same, but, surprisingly, both the pear and the apple contain some red colour and appear bright.

Red light

Looking at Figure 10, Figure 11 and Figure 12, the tomato is a very saturated red colour, this means that under red light it will look similar to what it looks under white light. Under green or blue light however, the tomato will appear very dark, because neither coloured light contains the red needed to highlight it.

The apple is yellow, a sum of red and green (see Figure 7), so it will appear bright under both red and green light, and dark under blue light. The pear is a muddy green colour, so it is visible in all pictures.

The consequence of this is that we can use a specifically tailored light source to emphasise the colour of a particular item of merchandise (Figure 13).

Figure 12: Still life lit by green light
Both the apple and the pear appear bright, while the tomato appears dark.

A similar application is the lighting of meats and produce in supermarkets: using a special light source the discolouration of the merchandise can be hidden, and the items on display made to appear fresher and more appetising. The spectrum of this type of lamp contains enough red to highlight the colour naturally present in the meat

On the other hand, a wrong decision can ruin the interior: excessive use of colours next to the tables of a restaurant for example would cause colour bleeding on the dishes staining the food and damaging the visual experience of the diners.

It is important to find a quantitative way of evaluating the ability of the lamp to reproduce colours in order to be able to compare the performance of two different lamps.

Figure 13: Highlighting items of merchandise using coloured light.
The blue of these trousers looks saturated by violet light. The disadvantage is that when the customer looks at them under white light, the actual colour will be different.

2.5 Colour rendering

The previously mentioned restaurant is one of those places where selecting fixtures that are able to let the customer appreciate every fine nuance of colour of the task is paramount. The performance indicator for this is colour rendering. It is defined[5] as follows:

> **Colour rendering** is the effect of an illuminant on the colour appearance of objects by conscious or subconscious comparison with their colour appearance under a reference illuminant. (5)

So the colour performance of the lamp in exam is compared to a reference illuminant (a special halogen lamp) in its ability to reproduce a set of eight colours for the average observer. The Colour Rendering Index (R_a) measures this ability to replicate the performance of the reference illuminant on the eight-colour sample. The index can be either a percentage, with perfect reproduction evaluated 100%, or a series of classes (from best to worst: 1A, 1B, 2, 3, 4).

The colour rendering index is a characteristic of the light source: for example halogen and incandescent lamps have an R_a index of 100%,

---
5. (EN 12665 2011)

Colour, colour temperature and colour rendering

while first generation linear fluorescent lamps had 65%. In my opinion however, one should consider the whole luminaire, instead of just the lamp: a coloured diffuser, for instance, would immediately ruin the performance of the lamp because it would introduce a dominant in the light and risk the effects shown in Figure 10 to Figure 12.

Figure 14: Comparison between light with very low (on the left) and very high colour rendering (on the right).

A clothing or accessories store, in particular one belonging to those brands that offer the same model of shirt, using the same textile and differing only for the colour, requires the highest R_a for two reasons: the obvious one is that colour is the selling point, the less obvious one is that restocking a shirt brought back by an unsatisfied customer has a cost.

R_a	Percentage	Example of activity
1A	90-100	Quality control
1B	80-89	Restaurants
2	60-79	Metal welding
3	40-59	Platforms
4	20-39	Parking lots

Table 4: Colour rendering: percentages, classes and example of activity.

A colour inspection laboratory for a textile manufacturing industry will also require a very high colour rendering index.

Figure 14 shows how much the appearance of a colour wheel changes when illuminated with a lamp that has poor colour rendering (left).

The European regulations for the lighting of indoor work places (EN 12464-1, 2011) specify the minimum required colour rendering index according to the different activities performed in the interior, some of which are listed as an example in Table 4.

Selecting a source with high colour rendering usually means selecting a slightly more expensive lamp, this is the reason why designers do not automatically go for the highest R_a. On the other hand it is very difficult to defend the choice of a really low quality light source (e.g. class 2 or 60 to 79%) because the awful colour rendering will ruin the well being in the interior.

2.6 Colour temperature

There is a striking visual difference between a warm light lamp and a cool one.

Table 5: Warm and cool light: its effects.

	Cool light	Warm light
Goes well with	Glass, steel, cool colours	Wood, leather, warm colours
Appearance	Brighter	Dimmer
Sensation	Cosiness	Stimulation

Warm light, such as candlelight, gives cosiness, it emphasises the warm hues of for example dark wood and leather, and the interior appears darker than it is.

Cool light, that from a wintry overcast sky for example, highlights cold hues, like those of steel and glass, and it makes the room look brighter than it really is (Table 5).

Colour temperature is measured in Kelvin (K), with higher numbers for cooler temperatures and vice versa. So for example the warm temperature of candlelight is around 1700 K, the cool temperature of the overcast sky is 6500 K.

Colour, colour temperature and colour rendering

The regulations define three groups of temperatures: warm (under 3300 K), neutral (between 3300 K and 5300 K) and cool (over 5300 K). So for example a lamp emitting cool light will have a colour temperature of at least 5300 K.

Strictly speaking, colour temperature (CT) is only for light sources that have a black body like behaviour (incandescent and halogen). For all the other sources the concept of Correlated Colour Temperature (CCT) is used, but, even though there is a difference in the definition, for all our purposes the two concepts are identical.

For completeness' sake I will introduce the definition of Correlated Colour Temperature (CCT) even though it refers to concepts that go beyond the scope of this book.

> **Correlated Colour Temperature** is the temperature of a Planckian radiator whose perceived colour most closely resembles that (6) of a given stimulus at the same brightness and under specified viewing conditions.

A Planckian radiator is somewhat similar to a white-hot piece of metal giving off light whose colour depends on the temperature of the metal itself, measured in Kelvin.

What this means is that the correlated colour temperature of a lamp that is not a halogen or incandescent is the temperature an halogen needs to have in order to look similar to it.

Figure 15: Example of lamp code.

Lamp manufacturers publish colour temperatures in their catalogues and websites, and print a code in all their products. Figure 15 shows a 58W Osram linear fluorescent lamp with the code 840: the first number is the CRI (8=80 to 89%), the second and third indicate CCT (40=4000K). Notice how the lamp says "Cool White", even if its colour temperature falls in the neutral group.

2.6.1 Reciprocal mega Kelvin

Let us consider a real life situation in which a designer has to choose between two lamps: one that has a CCT of 2700K, the other 3000K. He or she recalls that in another installation he used a 6500K lamp instead of a 6200K one, and the difference was not at all noticeable.

Will that remain true?

Unfortunately using only the Kelvin unit of measurement we have no way of finding the answer to this question. Experience will show that a small variation in Kelvin degrees will be very noticeable to the eyes if the temperatures are warm, and not so much if they are cool, but we need to introduce a quantifiable method of evaluating colour temperature differences.

Table 6: Colour temperature comparison between Kelvin and reciprocal mega-kelvin.

Definition	MK^{-1}	K
Warm	500	2000
	400	2500
Intermediate	300	3333
	200	5000
Cool	100	10000
	0	Infinite

In short, we need to introduce a different[6] unit that solves this problem: the reciprocal mega Kelvin (MK^{-1}). To convert a colour temperature from K to MK^{-1} you simply take the number one million and divide it by the Kelvins.

---

6. CUTTLE, LIGHTING BY DESIGN, 2008

The advantage of this system is that a difference of for example 100 MK^{-1} between two lamps means the same visual difference, no matter their actual colour temperature.

$$MK^{-1} = \frac{1\,000\,000}{K} \qquad (7)$$

Let us then try and apply the new unit of measurement. The calculations (Table 7) show that the difference between 2700K and 3000K is 37 MK^{-1}, while the difference between 6200 and 6500K is 7 MK^{-1}. This means that the former will be much more evident to the eye than the latter.

K	MK-1	Difference
2700	1000000 / 2700 = 370	
3000	1000000 / 3000 = 333	370 - 33 = 37
6200	1000000 / 6200 = 161	
6500	1000000 / 6500 = 154	161 - 154 = 7

Table 7: Variations of colour temperature: comparison between K and MK^{-1}.

There are no strict general guidelines or requirements about colour temperature, the regulations (EN 12464-1 2011) intervene only in a few specific cases where extremely high colour fidelity is necessary (in order to have the best colour perception, the CCT of the source must be from 4000K to 6500K). The increase in performance is limited and it only becomes necessary in the most demanding applications, for example precision work in arts and crafts, colour inspection and working with precious stones in jewellery manufacturing require an interval of 4000K to 6500K; examination rooms in healthcare premises need a narrower 4000K to 5000K; art rooms in art schools and colour inspection in multi-coloured printing require 5000K to 6500K.

In the 1940s a Dutch lamp engineer and physicist named Kruithof was writing a paper (KRUITHOF 1941) detailing the behaviour of the newly introduced fluorescent lamp, which, for the first time, allowed the designer to select the CCT for the light. In order to provide a guideline he presented the diagram that took his name, and is shown in Figure 16.

Figure 16: Kruithof diagram.
It links colour temperature, illuminance (a measure of light) and the sensation of well being in the interior.

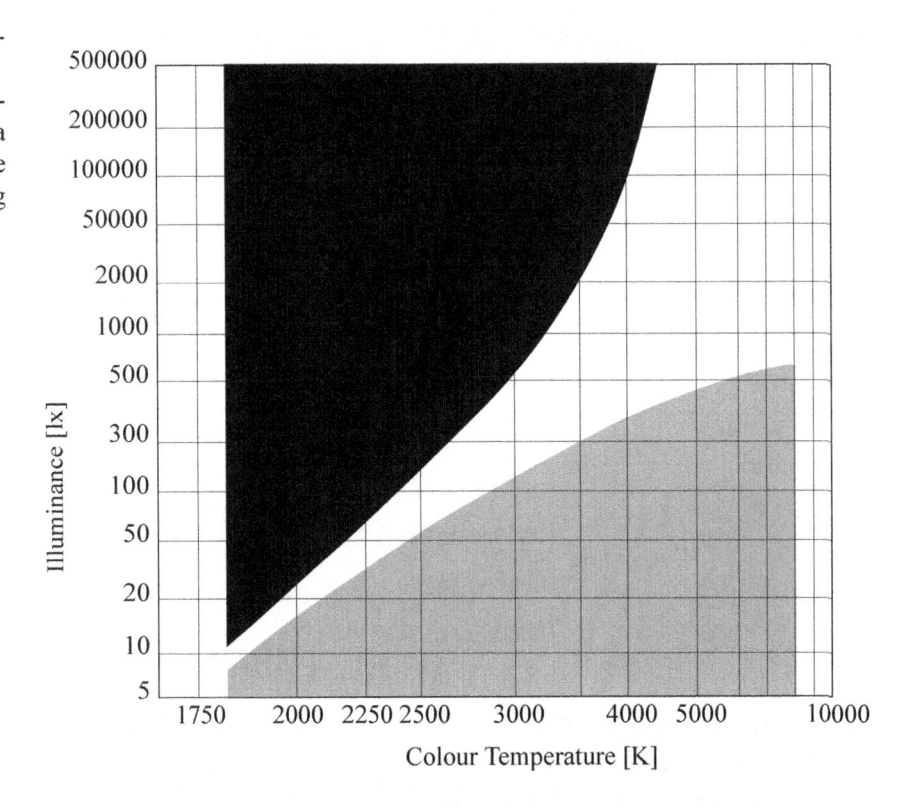

The diagram shows an empirical relation between high illuminance[7], high CCT and a feeling of well being. In the dark grey area one feels excessive warmth, and in the light grey area excessive cold. The area in the middle is where one feels that the levels of illuminance and the colour temperature of the light are pleasing. Even though he did not provide a full scientific backing, this curve has been proven more or less correct a number of times, and it has been part of good lighting practices ever since.

---

7. Amount of light on a task. Illuminance will be defined precisely in "6 Illuminance and evaluation of a lighting system" on page 61.

Colour, colour temperature and colour rendering

A very clever practical application of cool and warm light is the McCandless method for stage lighting (MCCANDLESS 1932), still very much in use. The objective is to find a way to represent shadow in an installation without actually creating areas with low luminance, which would be invisible to the audience and avoid the ridiculous situation one finds in some stores and exhibitions (see Figure 17) where total darkness is broken by narrow cones of light that highlight the items on display.

Figure 17: Installation without general lighting.

The installation should have two main lights, one coming from 45° to the right and the other from 45° to the left of the stage, to illuminate both sides of the actor. The clever idea is to use two different colour temperatures: cooler (or blue coloured light) on the shadow side and warmer (or yellow to orange coloured light) on the other. This increases the flow of light[8], while providing enough illuminance to perceive both sides of the scene.

---

8. Flow of light will be defined in 23.3 on page 292.

A similar trick is to use a cool colour for background lighting, in order to give the impression of darkness; and a warmer colour direct luminaire on the task, to highlight it. This will reduce luminance[9] contrast to manageable values, provide the general illuminance the user needs to move about in the interior and maintain the impression of darkness.

2.7 Self evaluation

Answer the following questions to test your understanding of the material:

1. Explain how the sensation of brightness changes according to the colour of the light.
2. Explain what are photopic and scotopic vision, and what are their properties.
3. Explain how the colour of a cyan object would appear if lit by blue light. What if it were lit by green light? What about red light.
4. Explain what is colour rendering, how it is measured and what it is for. Describe how objects will look both under a high and under a low colour rendering light.
5. Explain what is colour temperature, its properties, and the different ways in which it is measured, describing the advantages of each and the effects on the atmosphere.
6. What is the Kruithof diagram? What does it show?
7. What is the McCandless method for stage lighting?

---

9. Brightness. Luminance will be defined in 9.1 on page 97

Colour, colour temperature and colour rendering

Basic electrical quantities

3

This is the chance to define a few very basic concepts of electrical engineering that keep turning up here and there in an interior designer's job.

3.1 Electrical current

> ```
> Electrical current is the quantity of
> electrical charge (electrons) that flows
> through a conductor in an unit of time.
> It is measured in ampere (A).
> ```
> (8)

Frequently electrical quantities are compared to hydraulic ones for reasons of clarity. If the wire is similar to a pipe, the current is the volume of water that flows through it every second.

The lighting designer will need this when working with LEDs, as they often function at two different current levels (usually 350 mA and 700 mA), with the higher mode usually meaning higher output but shorter life.

3.2 Difference of electrical potential, voltage

> ```
> The difference of electrical potential, or
> voltage, measures the latent ability that
> the circuit has to produce work.
> It is measured in volt (V).
> ```
> (9)

It can be thought of as the difference, in height, between the top and the base of a vertical pipe full of water. The ability of the water to produce work will increase with the length of the vertical pipe, because the water will fall from a greater height i.e. with a higher potential energy.

An electric circuit functioning at a higher voltage, e.g. 220 V, will then be able to produce a lot more work than a low voltage circuit, e.g. 12 V. In the event of a short circuit, the former will be much more dangerous than the latter. This is why there is a regulation (IEC 60364) that, among many other things, compels the use of low voltage lighting fixtures (SELV) in places where there is a higher risk of shock, such as certain areas of bathrooms or swimming pools.

3.3 Electrical power

> `Electrical power` is the amount of energy that the circuit absorbs in a unit of time. (10) It is measured in watt (W).

It is similar to the amount of energy per second a pump will use to move water uphill. Both the size of the water flow (current) and the difference in height of the pipe (voltage) work against the pump.

Figure 18: Examples of commonly used lamps (not to scale) and the electrical power they absorb.
From left to right, on the first row we see an incandescent and a LED lamp, on the second row a halogen and a fluorescent tube, on the last row a compact fluorescent and a metal halide.

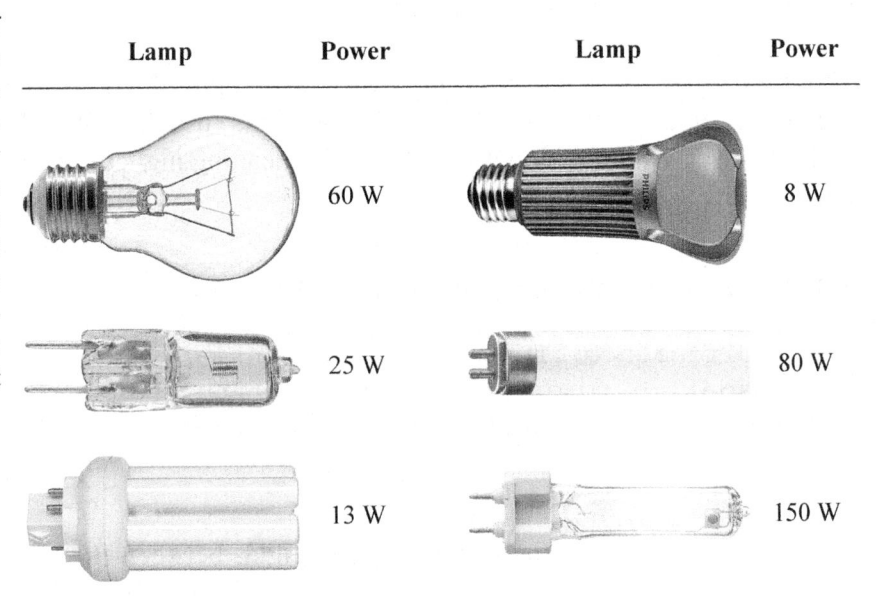

Lamp	Power	Lamp	Power
	60 W		8 W
	25 W		80 W
	13 W		150 W

Indoor light fixtures can go from just a few watt for LED strips to the hundreds of watt that halogen lamps absorb. Outdoor fixtures employ lamps that absorb up to thousands of watt.

Figure 18 shows different types of lamps and the electrical power they absorb.

A designer will use this concept when dimensioning the luminaire, because it shows how much electricity it will use.

Let's say that the ideal fixture for a specific task is a narrow beam spotlight. The designer chooses a 150 W one, and obtains three times as much illuminance[1] as he should have. The solution in this case is to find out whether there is a 50 W version of exactly the same light fitting, because that should produce the needed illuminance.

3.4 Self evaluation

Answer the following questions to test your understanding of the material:

1. As an interior designer, when should I use the concept of electrical current?
2. As an interior designer, when should I use the concept of voltage?
3. As an interior designer, when should I use the concept of electrical power?

---
1. Light on the task.

Luminous flux

4

What if the designer had to decide whether to employ a 70 W lamp of one type or a 150 W lamp of another? Electric power is only half the answer: it measures how much electricity the luminaire will consume, or how expensive the bill is going to be: e.g. a 100 W one will be half as expensive as a 200 W one. It says nothing about how much light is produced, and which lamp is the most efficient.

4.1 Definition of luminous flux

A lamp is a device that transforms electrical energy into something else: light, heat and UV. Let's look into each of these three emissions.

1. Heat can be transmitted from the luminaire in three ways: by conduction (the lamp is in contact with a cooler part of the fixture and heats it up), by convection (the lamp heats up the air which then heats up the body of the fixture), or irradiated (some of the radiation ejected by the lamp is IR). Heat is almost always wasted energy: the fixture must be designed to dissipate it, the luminaire must be placed in such a way that any risk of damaging the task or causing a fire is avoided, and the air conditioning system must counterbalance the rise in temperature caused by illumination.
2. UV radiation is very dangerous and must be filtered out.
3. Light is the bit in which we are interested. Obviously we would prefer a lamp that converts most of its electrical power into light.

The sensitivity of the eye changes with the wavelength (colour) (see Figure 6 on page 15). So we need to consider how the different colours that are produced by the lamp are transformed into sensation by the eye, and for this let's define luminous flux.

> **Luminous flux** is power emitted by the lamp in the visible spectrum weighted by the spectral sensitivity of the average human eye.
>
> It is measured in lumen (lm). (11)

We can now update Figure 18 on page 34 and add the luminous flux for each type of lamp. The result is in Figure 19.

Figure 19: Example of commonly used lamps (not to scale), electrical power they absorb and luminous flux they emit. From left to right, on the first row we see an incandescent and a LED lamp, on the second row a halogen and a fluorescent tube, on the last row a compact fluorescent and a metal halide.

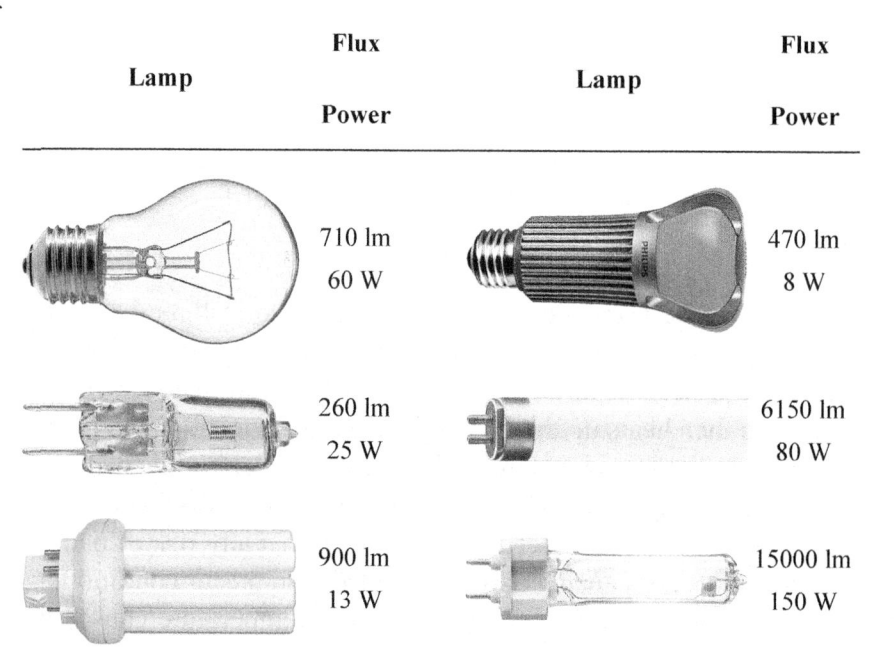

Lamp	Flux / Power	Lamp	Flux / Power
(incandescent)	710 lm / 60 W	(LED)	470 lm / 8 W
(halogen)	260 lm / 25 W	(fluorescent tube)	6150 lm / 80 W
(compact fluorescent)	900 lm / 13 W	(metal halide)	15000 lm / 150 W

4.2 Luminous efficacy

Lamps produce luminous flux and consume electrical power. This gives the designer a first criterion to select them: efficacy. The better lamp will be the one that produces the most luminous flux and absorbs the least electrical power.

Luminous flux

> **Luminous efficacy** (η) is the quotient of the luminous flux emitted by the power absorbed by the source (EN 12665 2011).
>
> It is measured in lumen over watt (lm/W). (12)

η[1] depends on the numerator (how efficiently the eye perceives the light that is produced) and on the denominator (how much power it took to produce it).

The value comes mostly from the physical process used: incandescence, for halogens and incandescent lamps, is a rather inefficient one because a large part of the energy will be turned into heat; while the electric arc used in metal halide lamps is more efficient. This means that a discharge lamp will have, on average, a higher efficacy than an incandescent one.

Luminous efficacy depends on the spectrum emitted by the lamp, because the eye's spectral sensitivity changes with wavelength, so for example, according to Table 1 on page 16, a monochromatic lamp that emitted[2] 20 W of power in red at 690 nm (sensitivity 0.0082) would emit a hundredth of the luminous flux of a lamp that emitted the same 20 W but in green at 530 nm (sensitivity 0.86).

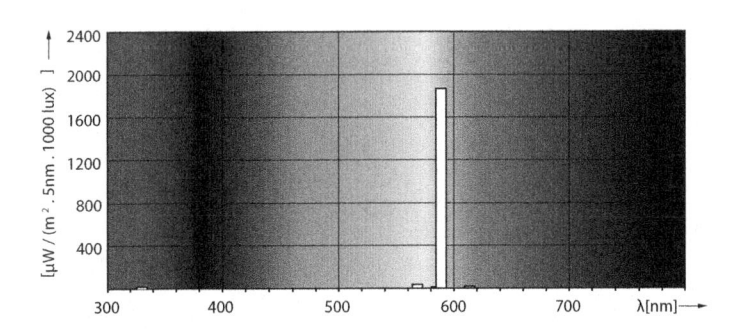

Figure 20: Spectrum of emission of a low pressure Sodium lamp (Philips Lightcolour SOX)

---

1. Greek letter eta.

2. The watt measures power emitted if it is radiometric (see "2.2 The observer" on page 14). If it is photometric, we use the lumen.

This means that the green light will appear a hundred times more luminous that the red, or that to provide the same sensation, the red light needs to be a hundred times more powerful than the green.

Using Figure 6 on page 15, we can determine what is the most efficient lamp imaginable and what will be its (ideal) efficacy. If the eye is most sensitive to a wavelength of 555 nm, then a lamp that emitted all its power at that exact wavelength would have the highest theoretical efficacy: 683 lm/W.

A real life example of this type of lamp was commonly used outdoors to illuminate parking lots and industrial areas: the low pressure sodium discharge lamp. An example of such a lighting system is shown in Figure 21. This lamp has next to no colour rendering: every colour turns into a darker or lighter orange. This causes a reduction of depth perception.

Figure 21: Outdoor area lit by sodium lamps.

It is not dissimilar from this monochromatic image, only the dominant is light orange., There is no colour rendering and a diminished depth perception. As a curiosity, the reader can see the spectral power distribution of the low-pressure version of this type of lamp in Figure 20.

Obviously such a low quality light has no useful application on any interior space that is not a parking lot, but it serves as a warning to the more colour-happy interior designers: too much coloured light might give a creative ambiance, but colour rendering is nil, and the interior is unappealing.

Lamp	Efficacy [lm/W]	Lamp	Flux
	12	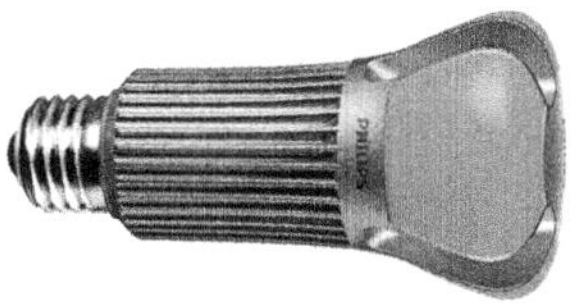	59
	10		77
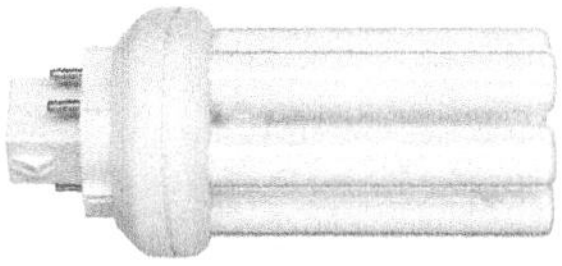	69		100

Figure 22: Examples of commonly used lamps (not to scale), with their luminous efficacy.
The incandescent and halogen have the lowest efficacy, next come the fluorescents and LED, and finally the metal halide.

We can see the lamps diagram updated with efficacy in Figure 22. We will look into each type of lamp in detail in part 2, but we can start drawing conclusions on the efficacy of the different types of sources. Halogen and incandescent lamps appear to be the least efficient of the lot with 10 and 12 lm/W respectively; the LED replacement for the incandescent has an average efficacy of 55 lm/W; the fluorescents have high efficacy at 69 to 77 lm/W; the best efficacy is reached by the metal halide, at 100 lm/W.

4.3 Light output ratio (R_{LO})

Similar to luminous efficacy, light output ratio evaluates the efficacy of luminaires instead of lamps. It measures the percentage of luminous flux that manages to exit the luminaire relative to the amount produced by the lamps.

A luminaire will absorb luminous flux for a variety of reasons that will be discussed in detail later in the book. For now let's say that the main causes of inefficiency are:

1. Reflectors[3] are not perfect mirrors: they may absorb up to 40% of the incident flux.
2. Refractors[4] are not perfectly transparent: they will usually absorb less than 5% of the incident flux.
3. Diffusers absorb about 30% of the flux.
4. A certain amount of flux will be lost into the body of the luminaire.

For more details about luminaires, please see also 17 on page 187.

> **Light output ratio** (R_{LO} or LOR) of a luminaire is the ratio of the total flux of the luminaire, measured under specified practical conditions with its own lamps and equipment, to the sum of the individual luminous fluxes of the same lamps when operated outside the luminaire with the same equipment, under specified conditions.
> It is a number with $0 \leq R_{LO} \leq 1$ (13)

The closer to 1 R_{LO} is, the most efficient the fixture will be. In real life we can expect the best R_{LO} to reach 80% to 85%.

---

3. Shaped mirrors that reflect light towards the visual task, usually made of aluminium.

4. Shaped transparent screens that refract light going though them.

Luminous flux

R_{LO} measures how transparent the luminaire is, and it contains the suppressed premise that being very transparent is a good thing. When we look into illuminance and into visual comfort, we will see why that it is not quite true. A qualitative example of this is looking at the sun: when direct sunlight enters an interior, it usually glares on the occupants; if on the other hand one puts a curtain on the windows then sunlight will be diffused and not so blindingly bright anymore. The curtain will act as a diffuser and absorb some of the light, but the visual comfort it guarantees more than makes up for the loss of luminous flux.

4.4 Total efficacy

It elaborates on the light output ratio concept (R_{LO}). Together with luminous efficacy, it is the main tool to evaluate how energy efficient the fixtures are.

In particular, the combination of the two concepts is rather interesting:

> `Total efficacy` is the product of light output ratio and luminous efficacy. (14)

It shows how efficiently the light fitting is emitting luminous flux. It is the number of lumen emitted from a fitting per each watt absorbed.

As an example let's consider two luminaires: one with $R_{LO} = 57\%$ and a lamp with $\eta = 55$ lm/W, and another with $R_{LO} = 65\%$ and a lamp with $\eta = 45$ lm/W. Since R_{LO} is very easy to calculate, and is commonly used as a marketing tool while η needs to be looked up, the uninformed designer might stop at light output ratio and believe that luminaire 2 is more efficient.

Let's calculate TE instead.

> `TE1` = $R_{LO}1$ • η = `57%` • `55 lm/W` = **`31 lm/W`** (15)
> `TE2` = $R_{LO}2$ • η = `65%` • `45 lm/W` = **`29 lm/W`** (16)

Looking at (15) and (16) shows that it is actually the other way around.

4.5 Luminaire's luminous efficacy (LLE)

The advantage of total efficacy is that it takes into account both the luminaire and the lamp. The limit is that it does not contain any information on the efficacy of the electromagnetic or electronic devices that the luminaire needs to work (see 11.2.6 on page 131). In the case of a LED for example, this can be quite significant. The LED in the spotlight used in Figure 244 on page 486 absorbs 6 W, while the whole luminaire absorbs 9 W. On an emitted flux of 570 lm, the difference is significant. Considering only the LED, one would get an efficacy of 95 lm/W, but when the power absorbed by the whole luminaire is counted, it turns into 63 lm/W, with a reduction of 44%!

It is the translation of the same concept seen in 4.2 to the whole luminaire.

> `Luminaire's luminous efficacy` (`LLE`) `is the quotient of the luminous flux emitted by the power absorbed by the luminaire.` (17)
> `It is measured in lumen over watt (lm/W).`

This indicator gives a complete picture of the efficacy of the luminaire as a light-manufacturing engine.

The limit of this concept is that it does not contain any information about what happens to the light after it gets out of the luminaire. If we have to illuminate a painting hanging from a wall and we employ the best luminaire that total efficacy can find, we have no assurance that the light fitting will actually illuminate the task.

A more powerful indicator that surpasses this limitation is the Specific connected load, defined in (60) on page 264, but it requires concepts not yet introduced.

4.6 Self evaluation

Answer the following questions to test your understanding of the material:

1. What is the difference between electrical power and luminous flux?
2. Why was it necessary for us to introduce the concept of luminous flux?
3. Define luminous efficacy, explain what it is for and what are its limits.
4. Explain the concept of light output ratio, its function and its limit.
5. Explain what is total efficacy, what is it for, and why it is limited.
6. Explain the difference between luminous efficacy of the lamp and of the luminaire.

Luminous intensity

5

Now that we know how much light is produced by the lamp (Luminous flux) and how well (Luminous efficacy), we need to start looking into where it goes, and to do that I must first refer to two familiar concepts.

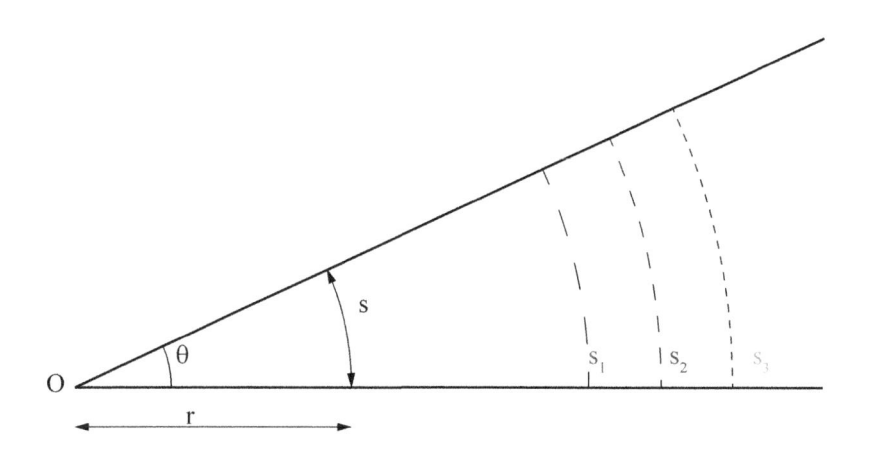

Figure 23: Plane angle, definition and angular size of arcs.

Let's look at Figure 23.

> A **plane angle** (θ) can be defined as the quotient of the length of the arc of circumference subtended (s), by the length of the radius (r).
> This also measures its size in units of radians (rad). (18)

The whole circle measures 2π radians, half a circle is π, and a quarter of a circle is $\pi/2$.

On the right side of Figure 23, multiple arcs of circumference s1, s2, s3 grow in length as one goes farther from the centre O. At the same time, to an observer placed in O and looking towards s_1, the angular sizes of s_1, s_2 and s_3 are the same, because as the arc grows, so does its corresponding radius r in such a way that the angle, which is the quotient between the two, stays the same.

The solid angle extends this concept to space (Figure 24).

Figure 24: Solid angle.

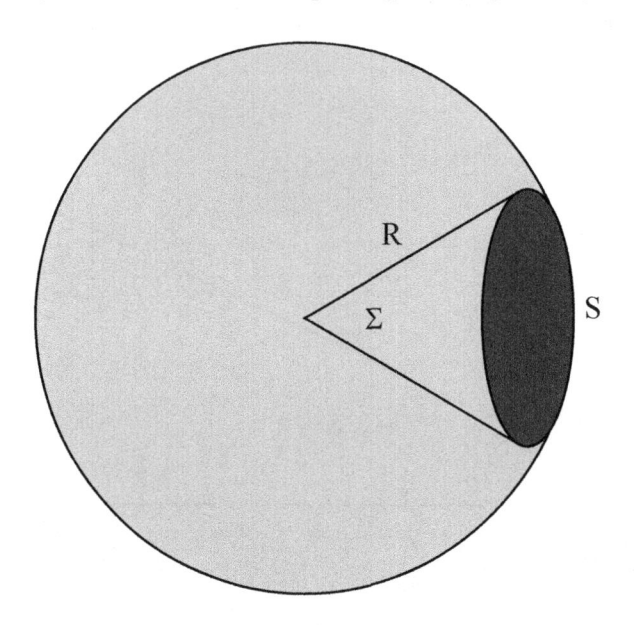

The definition follows:

> A **solid angle** (Σ) can be defined as the quotient of the surface of the segment of the sphere subtended (S), by the square of the radius (R).
>
> This also measures its size in units of steradians (sr).

(19)

The whole sphere is 4π steradians, half a sphere is 2π, a quarter of a sphere is π and so on.

The solid angle is a way to measure how large a task appears to the observer. This depends both on the real size of the task and on the distance from the eye, so that an object very far away could seem to

Luminous intensity

be the same size than a smaller one closer. As an example, consider that you may blot out the sun with a coin, if it is close enough to your face, even though your coin is much smaller than the sun. This happens because your hand is much closer to your eye than the sun, so the solid angle they cover can be the same.

In lighting, this means that the cone of light emitted by a luminaire (e.g. a spotlight) can illuminate either a small task close or a large one far away.

As I move the coin farther away from my eye, the solid angle will become smaller and smaller until it becomes so small as makes no difference to a direction in space (Figure 25). The luminous flux that goes in that direction divided by the solid angle is luminous intensity.

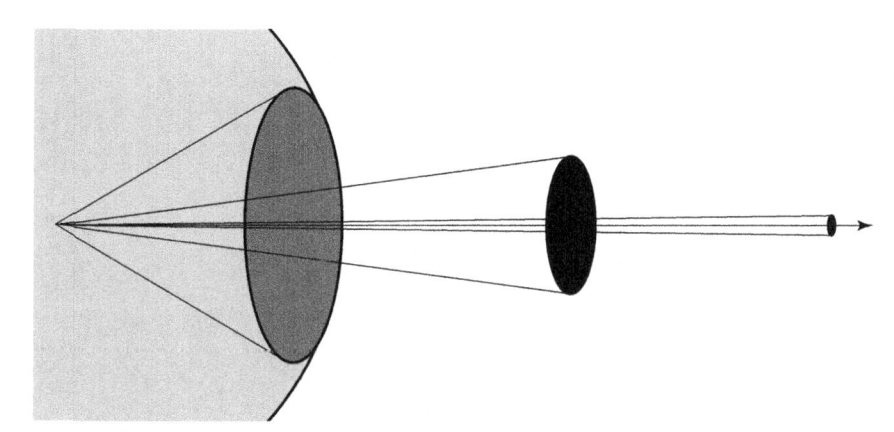

Figure 25: A small solid angle becomes a direction.

> `Luminous intensity` (I), of a source in a given direction, is the quotient of the luminous flux dΦ leaving the source and propagated in the element of solid angle dΩ containing the given direction, by the element of solid angle (EN 12665, 2011). (20)
>
> It is measured in candela (cd).

The luminous intensity tells the designer how much perceived light is emitted by the luminaire in that specific direction.

In fact a function of the luminaire is to shape the luminous flux produced by the source in the directions where the task is. This means that if we find a way to see the intensities emitted by the luminaire we can evaluate its performance and find the right fixture for our task.

5.1 Photometric solid

If we plot the intensities coming out of a luminaire we get a sort of hedgehog of vectors coming out of the centre of the fixture. When we join the tip of each of the arrows, we get a potato-like solid. Starting from the centre of the fixture, the intensities grow in the direction of more potato.

Figure 26: Photometric solid for a wall-washer luminaire.

Figure 26 shows the photometric solid for a ceiling mounted wall-washer: there is no intensity emitted over the horizontal plane (the fixture is recessed), and the arrow indicates the highest intensity (most potato).

These three dimensional light intensity distributions are not very practical. The best thing to do is to define a set of planes and cut the potato along those planes to obtain a series of curves.

Luminous intensity

5.2 C, gamma system of reference

In order to be able to compare different fixtures we need to set a standard system of reference for sectioning the photometric solid. The regulation (EN 13032-1, 2004) describes multiple systems; I selected the most commonly used one: the C, γ.

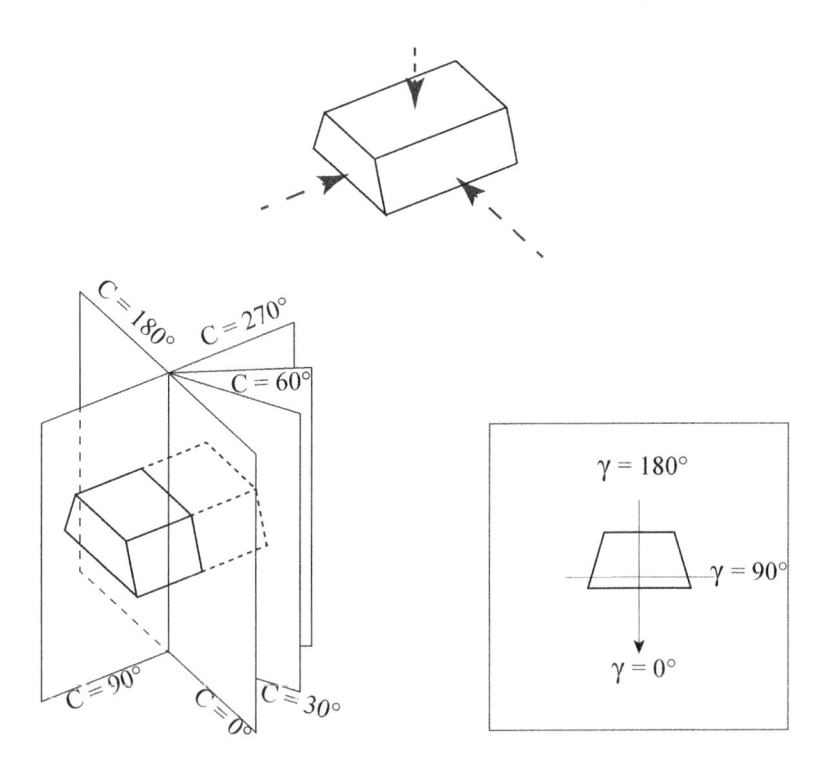

Figure 27: C, γ coordinate system. Optical centre of the luminaire (top). C planes (bottom left); gamma angles (bottom right).

First we define the optical centre of the fixture: for simplicity let's say it is the centre of the lamp[1], and there we place the origin of our system of reference. We define (Figure 27) the first axis as the one perpendicular to the light emitting area of the luminaire. We then

---

1. It is actually the point in a luminaire or lamp from which the photometric distance law operates most closely in the direction of maximum intensity.

define a second axis according to the type of luminaire: again for simplicity let's say it is perpendicular to the lamp's axis[2].

We can then define a series of half planes that share the first axis and slice the fixture. Each of them is designated as C_X with $0° \leq X \leq 180°$ and X indicating the angular distance from the second axis. Within each C plane the direction is given by a γ angle with $0° \leq γ \leq 180°$ that measures the angular distance from the first axis.

Figure 28: Shape of the photometric solid and light intensity distribution curve.

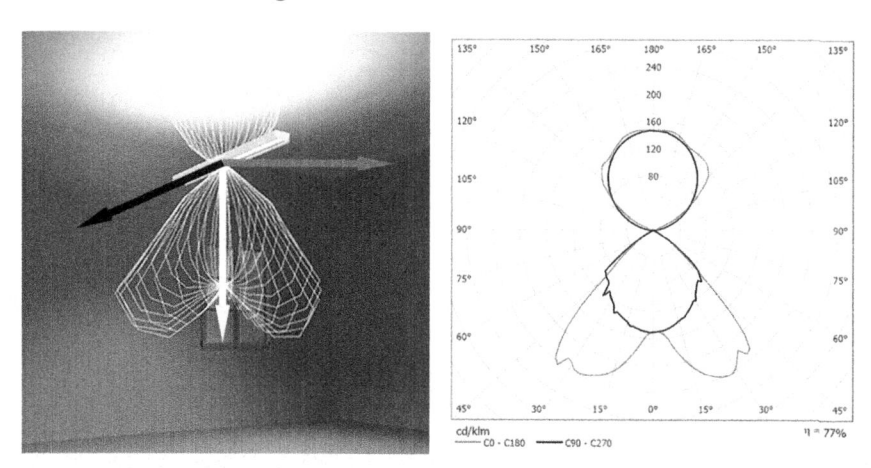

For example, in the left part of Figure 28 we have the white axis as the first axis, perpendicular to the light emitting area, the grey axis as the second axis, perpendicular to the lamp, and the black axis as the third axis. The grey curve in the right side of the picture is on the plane identified by the white and grey axis, and the black curve in the plane described by the white and black axis.

For simplicity opposite C_X half planes are usually displayed together to form a single plane called $C_{X-X+180°}$, e.g. $C_{0°-180°}$, which is the plane crosswise to the lamp axis, or $C_{90°-270°}$ (the lengthwise one).

---

2. It actually changes with the type of luminaire. For a luminaire with a single or double ended lamp not aligned with the first axis, the axis of the lamps is the third axis. For a tilting luminaire, the second axis is parallel to the tilting axis. For other luminaires, the maximum intensity must be in plane C0 at gamma zero, and the plane C0-180 is the plane of maximum symmetry.

5.3 Light intensity distribution curve

Figure 29 shows a luminaire with a double fluorescent lamp and its light intensity emission solid.

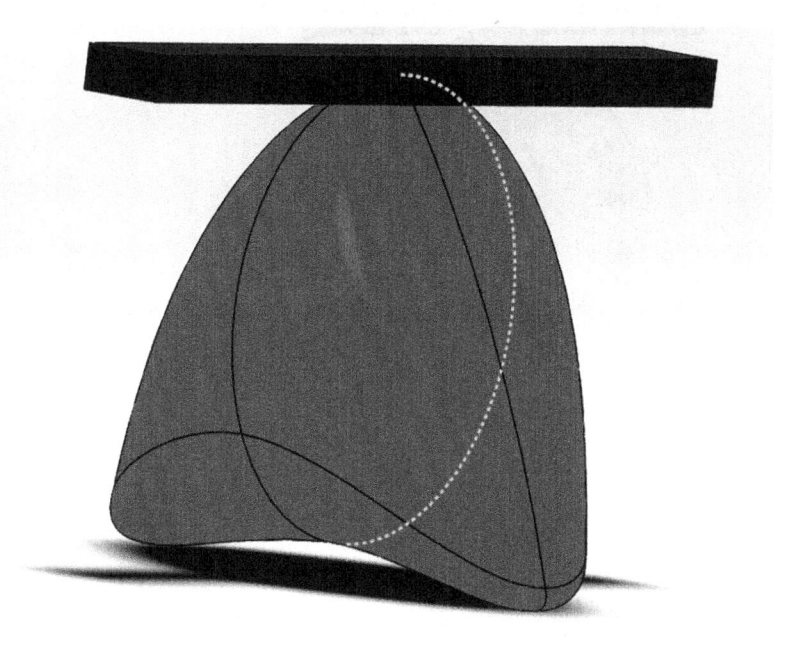

Figure 29: Fluorescent luminaire with emission solid.

Slicing the emission solid crosswise from the lamp produces the $C_{0°-180°}$ plane (Figure 30).

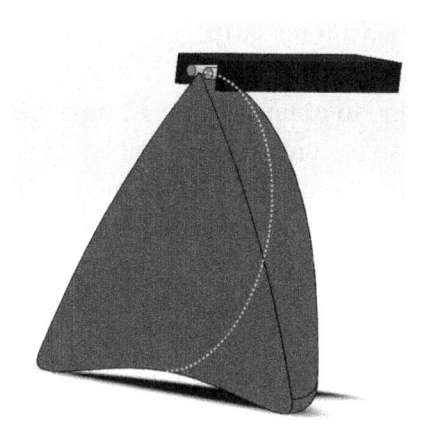

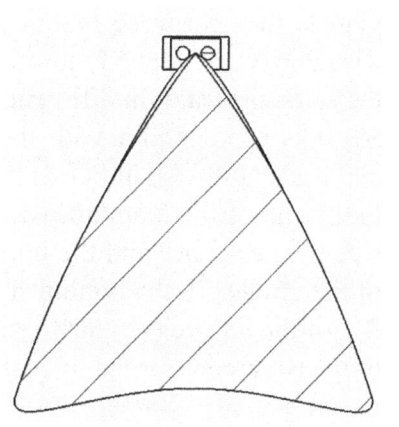

Figure 30: Crosswise section of the emission solid: $C_{0°-180°}$ plane.

Slicing lengthwise from the lamp produces the $C_{90°-270°}$ plane (Figure 31).

Figure 31: Lengthwise section of the emission solid: $C_{90°-270°}$ plane.

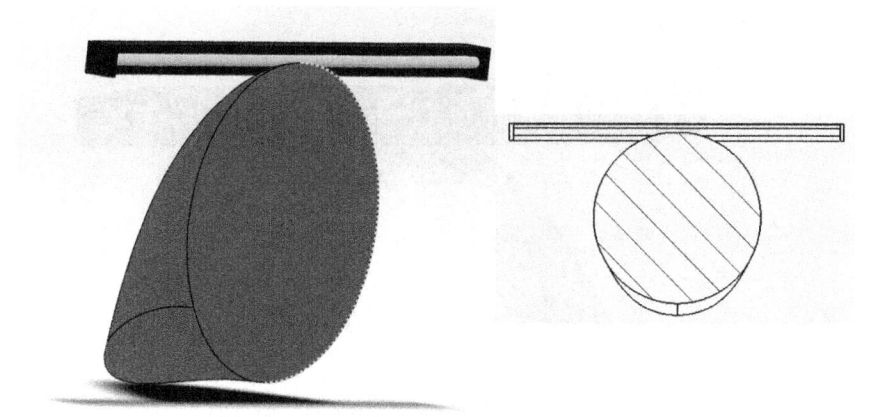

The result of cutting the photometric solid with a plane is a photometric curve, also called light intensity distribution curve.

> The **distribution of luminous intensity** is the display, by means of curves or tables, of the values of the luminous intensity of a source as a function of direction in a plane (EN 12665, 2011). (21)

The photometric curves normally given are the ones from two planes: the crosswise one ($C_{0°-180°}$, see Figure 30) and the lengthwise one ($C_{90°-270°}$, see Figure 31). Both curves are usually drawn on the same diagram, in different colours, to obtain Figure 32 on page 56. It is a polar plot with the crosswise curve displayed in grey and the lengthwise in black; the various γ angles are indicated near the frame of the diagram, with $\gamma = 0°$ in the middle of the bottom side of the square and the intensities are proportional to the radius of the circles. In the bottom right corner of the diagram there is the R_{LO} of the fixture ($\eta = 58\%$), and on the bottom left there is the unit of measurement for the intensities: it is candle over kilolumen.

This might surprise the attentive reader, because I wrote earlier that the unit of measurement for intensity is the candle, while here we see an intensity-over-flux unit. The reason for this is best explained

Luminous intensity

with an example. Suppose a luminaire manufacturer decided to employ two versions of the same lamp, with different powers, in the same fixture. This means that he can use the exact same set of components – in particular the same injection moulds (very expensive) for both fixtures, reducing the industrial cost per item. Both lamps have the exact same shape but emit a different amount of luminous flux. Since the shape of the two fixtures and of the two lamps are exactly the same, the fixtures will emit intensities proportional to the luminous flux emitted by the lamps. This means that the two fixtures will have light intensity distribution curves that look exactly the same, but for the scale. In order to avoid plotting two identical curves that differ only for the scale, manufacturers divide the intensities by the lamp's flux and plot the result.

The work-flow to obtain the actual luminous intensity from one of these diagrams is:

1. Find out the luminous flux emitted by the lamp(s) installed (e.g. 6000 lm or 6 klm). This is usually available in the catalogue of the fixture's manufacturer, or, failing that, in the lamp catalogue.
2. Read the intensity over flux in the diagram for the direction needed (e.g. for C_{0-180} and $\gamma = 30°$ we read 240 cd/klm).
3. Multiply this value by the sum of the flux emitted by each lamp, obtaining the actual intensity (in the example we multiply 240 by 6 and get 1440 cd).

Light intensity distribution curves can be plotted in Cartesian coordinates as well, even if this is a less common way, in particular because it makes it harder to evaluate how good the curve is[3].

Observe that all light intensity distribution curves must meet at $\gamma = 0°$ and $\gamma = 180°$ (see Figure 32).

---

3. We shall see why in "7.4 Illuminance uniformity and light intensity distribution" on page 86

Figure 32: Use of a light intensity distribution curve to find out the value of the intensity in a specific direction.

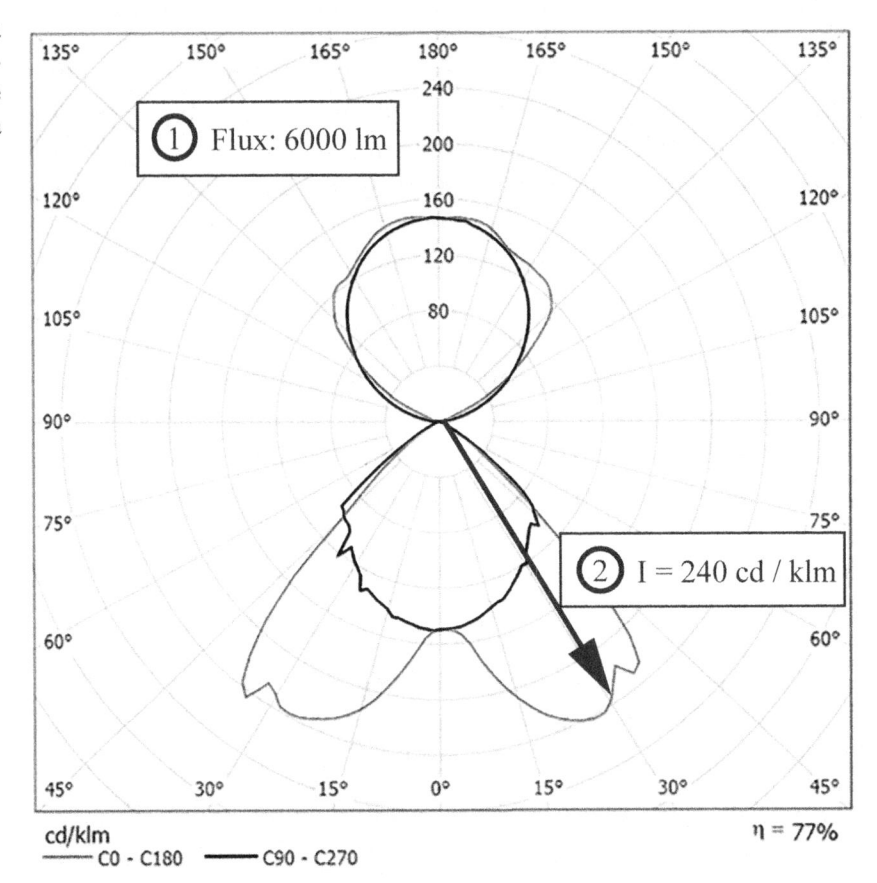

5.4 Beam spread

Luminous intensity is the tool to evaluate and dimension a luminaire when it is used in conjunction with another concept I haven't introduced yet (see "7 Intensity and illuminance" on page 79 and "6 Illuminance and evaluation of a lighting system" on page 61). On its own it is more limited, but it can help in dimensioning spotlights (see "19.5 How to dimension an axially symmetric luminaire" on page 229).

When selecting a spotlight, we might think that the total width of the light intensity distribution curve should be used So, in Figure 33 one might expect to use all the γ angles from - 75 ° to 75 °. The problem is that we would get a lot of intensity in the centre of the task and

Luminous intensity

next to none near the border, because the intensity at $\gamma = 0°$ is around 250 cd/klm and at $\gamma = 75°$ it is zero.

The concept of beam spread comes to the rescue in this case:

> **Beam spread** is the angular distance described by intensities larger than half the maximum. (22)

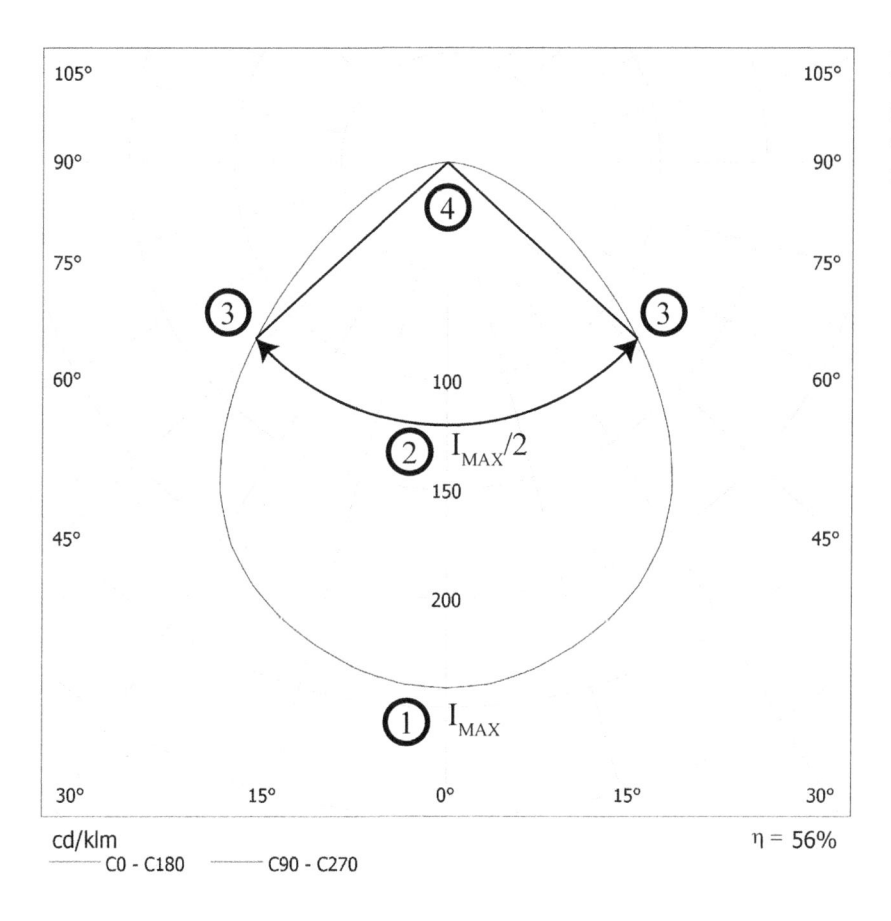

Figure 33: How to find the beam spread of a luminaire graphically, according to the definition in (22).

The beam spread shows what part of the intensity emitted by a spotlight is considered useful, and how different it is from the total width of the light intensity distribution.

Looking at Figure 33, this is the work flow:

1. Measure the maximum intensity (I_{MAX})
2. Halve that number ($I_{MAX} / 2$)
3. Find the points in the light intensity distribution curve that emit that intensity
4. Measure the angle between them.

Figure 34: Effect of a spotlight with too narrow a beam spread.

In the example, the beam spread is about 90°.

This means that the centre of the task will receive a luminous intensity equal to I_{MAX}, and its border half the maximum intensity, so the task will have greater uniformity.

If we selected a fixture with a smaller angle, we would obtain the result in Figure 34, if we selected a wider one we would end up illuminating the floor (which is not a task) thus wasting energy.

It is important to note that the beam spread is only partially responsible for the size of the bright spot. The brightness of the background is the main cause of the size of the spot: if it is very high, such as during a sunny summer day outdoors, the bright spot of the artificial light might not even be visible; while at night it will be about as large as it can be.

5.5 Self evaluation

Answer the following questions to test your understanding of the material:

1. Define the concept of solid angle and explain its use as an interior designer.
2. Define luminous intensity and explain its applications.
3. Describe the C, γ system of coordinates.
4. Define and explain the applications of the photometric curve.
5. Explain the concept of beam spread and its use.

Illuminance and evaluation of a lighting system

6.1 Visual task

It is interesting to examine the definition of visual task because it expresses the dual nature of a lighting system: freedom and necessity.

There is a part of the project in which the designer is completely free to operate and to express his professional creativity. For example the decision whether a wall will be a visual task or not depends on what exactly happens on it. The designer is free to decide whether it will be just a wall, or whether it will host some content (shapes, finishing, colours and so on) that makes it relevant to the interior. In the former case, it will not be a visual task, in the latter it will.

There is another part of the project that is governed by necessity. For example, the steps on a staircase need to be a visual task, because the users of the interior need to know where they are to avoid injury. The same goes for the floor of an interior. The designer is not totally free to decide how much illuminance the floor will receive because there is an absolute minimum that is necessary to allow people to actually use the place.

The regulations detail exactly this: they provide minimum requirements that are necessary if a certain activity needs to be performed.

> A **visual task** is a surface that is important because either an activity is performed, or something is shown in it that is relevant for the success of the installation. (23)

6.2 Task area

The reader is now familiar with two quantities that can work as indicators of performance: luminous flux and intensity; the former describes the light source; the latter is applied to luminaires.

In the next paragraph we can finally introduce an indicator that deals with our objective: illuminating something. This something is usually a task area, defined by CIE as follows:

> **Task area:** it is the partial area of the work place in which the visual task is carried out. (24)

6.3 Concept of illuminance

We must now find out what happens to a task surface when it receives luminous flux. Right now we don't care from where it is coming or how it was produced, we only care about the interaction of the flux with the surface.

To do that, we need to introduce the concept of illuminance.

> **Illuminance** (E) at a point in a surface is the quotient of the luminous flux dΦ incident on an element of the surface containing the point, by the area dA of that element.
>
> It is measured in lux (lx): one lux is one lumen of luminous flux over a square metre of surface. (25)

Illuminance is the amount of light that can be seen by the average eye spread onto a unit of the task surface. Given the surface, the more luminous flux, the more illuminance on it. Conversely, given the luminous flux, the larger the surface, the smaller the illuminance.

A spotlight, for example, will be able to produce a large amount of illuminance on a small surface that is close to it, or a small amount of illuminance on a large surface far away.

Table 8 shows the illuminance levels one can expect on the floor in familiar circumstances.

Illuminance of daily situations	
Bright sunlight	100000 lx
Office	300 lx
Parking lot	20 lx
Full moon	0.15 lx
Starry sky	0.002 lx

Table 8: Levels of illuminance in familiar circumstances.

The numbers give us a quantitative measure of the amount of light that ends up on a surface. Let's see what this means using an example. Suppose a person went to open his car after a night out and dropped his keys. If the car were parked in an area with no artificial lighting and in a moonless night (under starry sky: 0.002 lx), the process of finding the lost keys would be rather tedious and might entail quite a bit of rustling on the ground searching by touch. If the car were instead parked in a well-lit parking lot (20 lx), the search would be visual and much easier. So the activity "searching for lost car keys while grumbling" can be easily performed at 20 lx, it can't at 0.002 lx.

In other words, the illuminance level on a task allows certain activities and impedes others. A good rule of thumb is: the more complex the activities, the higher the illuminance required.

European regulations use illuminance as a requirement: there are tables (EN 12464-1, 2011 for indoor, and EN 12464-2, 2014 for outdoor tasks) that link each working activity to a prescribed minimum average maintained illuminance on the task areas.

Table 8 can tell us something else: we observe that there is a major difference between what is required in an office (300 lx) and the illuminance of sunlight (100000 lx), but we know that the performance of our eyes is similar. This is why a lighting system must not try and emulate the illuminance daylight produces, because it would mean a colossal increase in cost with maybe a minor improvement in performance for the observer.

It emphasises also how adaptable the human eye is: we can see well enough to negotiate our surroundings on a full moon at night (0.15 lx), and to function perfectly on a sunny day (over 100000 lx). This means that our eyes work in a staggering range of six orders of magnitude. If our muscles had that range, we would be able to lift a ping pong ball or two cars with the same ease.

6.4 Vertical and horizontal illuminance

We can distinguish multiple types of illuminance according to the absolute position of the task surface and its shape.

Vertical illuminance is needed when the surface of the task is flat and vertical, like a painting (for more details see "19.3.3 Application: horizontal illuminance" on page 218).

Horizontal illuminance lights any flat and horizontal surface, such as a carpet, a floor or a desk(for more details see "19.3.4 Application: vertical illuminance" on page 219).

6.5 Cylindrical illuminance

Cylindrical illuminance is necessary when the surface of the task is cylindrical and vertical, like the outside of a pole.

Setting a certain level of cylindrical illuminance is equivalent to requiring the system to provide illuminance on a vertical plane no matter its orientation in space. If vertical illuminance is the right choice to highlight for example a painting on a wall, because there is a preferred direction for viewing it, a rugby player in an arena requires cylindrical illuminance, because the fans surround the player.

In an interior it is a useful tool for evaluating visual communication, because it allows people to recognise each other's features, and any vertical bit of information on display.

For more details see "19.3.6 Application: cylindrical illuminance" on page 224.

6.6 Semi-cylindrical illuminance

Semi-cylindrical illuminance considers only half of the lateral surface of a cylinder, and it is very important in lighting to illuminate the human face and figure.

Again it is similar to requiring a certain level of vertical illuminance in multiple directions, this time around a main one. In the case of a person's face, the main direction is towards the nose; for a statue of a person standing up, placed next to a wall in a church for example, the main direction of view is perpendicular to the wall.

For more details see "19.3.5 Application: semi-cylindrical illuminance" on page 223.

6.7 Evaluation of a lighting system: performance

I can now introduce a mean of describing how easy it will be to perceive objects in the interior (average maintained cylindrical illuminance and uniformity) and how pleasant the illumination is going to be (modelling). I will then talk about the most powerful tool for evaluating the lighting system: the isolux curves.

Let us begin by looking at what happens to a lighting installation as time goes by.

6.8 Maintenance factor

Luminaires do not emit their initial luminous flux for all their lifetime: the lamps grow old, the reflective, refractive and diffusing surfaces in the fixture become dim, and the environment has an effect on the performance: dust and dirt accumulates. This means that the task surface will receive less flux hence a lower illuminance.

If we designed the system so that it complied with the regulation when everything is new, all these factors would quickly reduce its illuminance, lose the compliance and make it harder or impossible to perform the required activities.

In order to avoid that, we use the maintenance factor (see 6).

> **Maintenance factor** (MF): ratio of the average illuminance E_t, on the working plane, after a certain period of use, of a lighting installation to the initial average illuminance E_0, obtained under the same conditions for the installation (EN 12665, 2011).
> $0 \leq MF < 1$ (26)

The idea here (see Figure 35) is to measure illuminance after a certain amount of time, and evaluate how much smaller in percentage it is compared to the initial value, then over-dimension the initial illuminance by that same factor.

The maintenance factor is a number often varying from 0.45 to 1, the higher the number, the cleaner the installation. For clean interiors the most commonly used value is 0.8.

When the maintenance factor is taken into account, illuminance acquires the sub "m", so if average illuminance is $\bar{E}$, average maintained illuminance is $\bar{E}_m$.

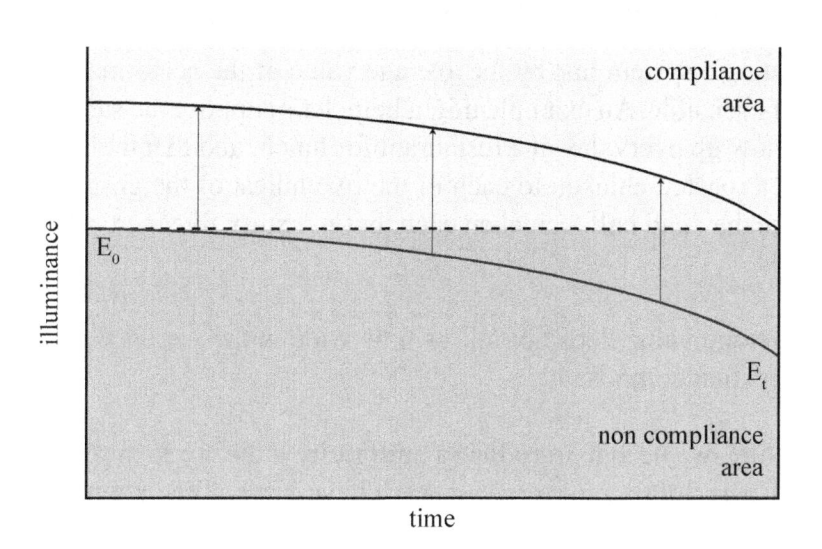

Illuminance and evaluation of a lighting system

Figure 35: Graphical explanation of the maintenance factor.

6.9 Average maintained cylindrical illuminance ($\bar{E}_{Zm}$)

As I wrote in 6.5 on page 64, the advantage of cylindrical illuminance is that it lets the viewer perceive details on a vertical surface no matter its orientation. Thus it is the perfect tool to evaluate the performance of the installation in visual communication and recognition of objects.

The regulations (EN 12464-1, 2011) require that in the activity and interior areas the maintained mean cylindrical illuminance must be no less than fifty lux. In areas where good visual communication is important (for example offices, meeting and teaching areas) the minimum requirement rises to a hundred and fifty lux.

The regulations have a requirement for uniformity as well, which will be discussed in the next paragraph.

6.10 Illuminance uniformity (U_0)

Evaluating a system just by the average value of the performance is not very sensible. An example might help: let's suppose the same ten men show up every day at a restaurant for lunch, and that the waiter serves a roasted chicken to each of the five tallest of the group. On average they had half a chicken each, but in reality five men eat and five fast.

The average value does not tell us how wide is the variance of the numbers that compose it.

In lighting, we need to introduce a uniformity indicator to make sure that e.g. the 300 lx on average that we have in an office are not the result of 2 lx areas interspersed with spots at 50000 lx. If this happened, it would be very hard to use the interior, and we would have serious problems with visual comfort[1].

> **Illuminance uniformity** U_0 is the quotient of minimum illuminance by average illuminance on a surface (EN 12665, 2011).
>
> $0 \leq U_0 \leq 1$ with the best uniformity at 1. (27)

This definition can be applied to luminance as well.

U_0 measures how close the minimum and the average are, and consequently the possible range for the values of illuminance in the various points of the interior.

When U_0 is close to one (Figure 36), it means that the minimum and average illuminance are close. If this is true, then the values between minimum and average will also be close (notice that the curve in the figure is rather flat) so there will be good uniformity in the area.

---

1. Visual comfort will be defined in "9.3 Visual comfort" on page 102.

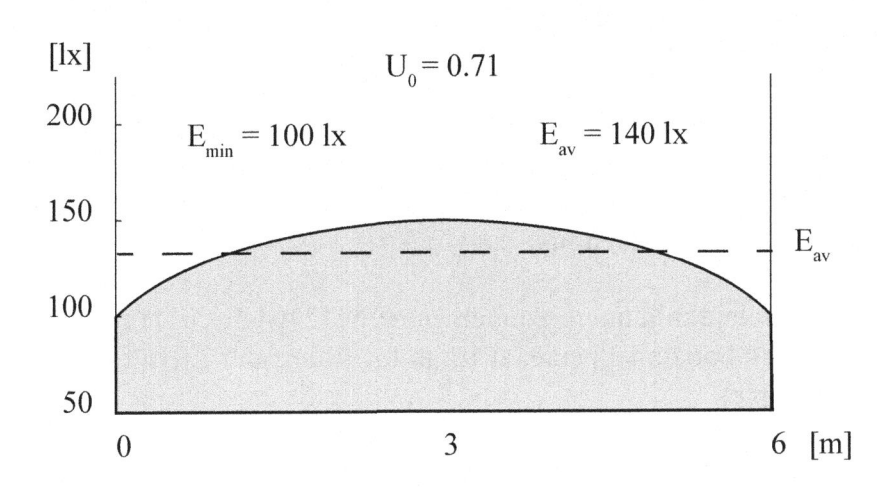

Illuminance and evaluation of a lighting system

Figure 36: Uniformity close to 1: minimum and average close to each other.

[Graph: $U_0 = 0.71$, $E_{min} = 100$ lx, $E_{av} = 140$ lx; curve spanning 0 to 6 m, y-axis in lx.]

When U_0 is close to zero, it means that the average and the minimum are very different from each other, like in Figure 37.

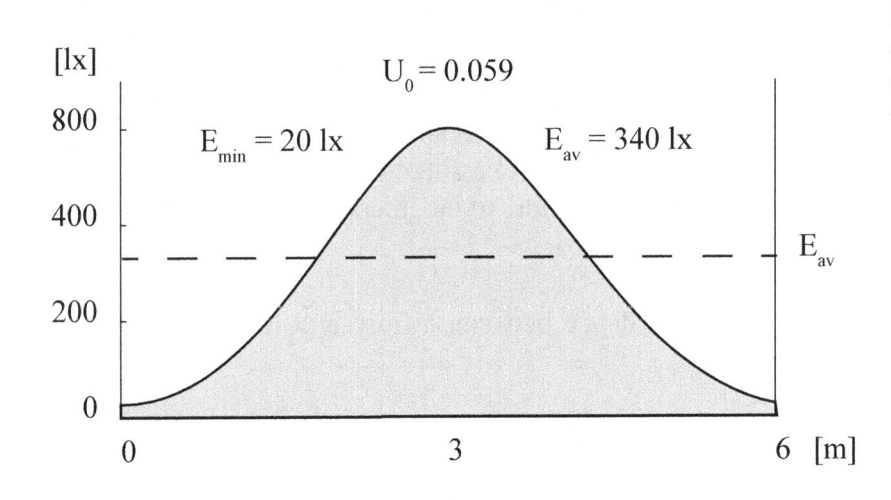

Figure 37: Uniformity close to 0: minimum and average far from each other.

[Graph: $U_0 = 0.059$, $E_{min} = 20$ lx, $E_{av} = 340$ lx; bell-shaped curve spanning 0 to 6 m, y-axis in lx.]

Consequently, the values in between will span a larger range, and the curve in the figure will be more like a hill than a plain, so there is less uniformity.

To see how powerful a tool uniformity is, let's look at an example. Let's compare two proposals for an office five metres by four, one producing $\bar{E}_m$ of 331 lx and a U_0 of 0.643, the other $\bar{E}_m$ of 357 lx and U_0 of 0.134. If we considered only the average maintained illuminance we might believe that the former proposal is worse than the latter, as it produces twenty-six less lux.

Only when we add U_0 the picture is complete: the first proposal produces a much better uniformity than the second, at a cost of only twenty-six lux, while maintaining an average well over the requirement of three hundred lux.

The absolute minimum requirement (EN 12464-1, 2011) for U_0 is 0.10 on a horizontal plane (at 1,2 m for sitting and 1,6 m for standing people).

Looking deeper in the definition, we can say that uniformity U_0 tells us whether the fitting is capable of illuminating the task well: that is how badly lit is the area that receives the least amount of luminous flux (minimum illuminance) compared to the average.

6.11 Modelling

Modelling (EN 12464-1 2011) is an indicator of how pleasant lighting is, or how close we are to a situation where the people, objects, textures and general structure of the interior are revealed clearly and pleasurably.

It describes the balance between diffuse and directed light, or between:

1. Excessive directional lighting: which casts very stark and deep shadows; it highlights any imperfection and any roughness in the surface finishing of the objects; it creates a very theatrical effect (Figure 38).

2. Multiple shadows: derive from a multiplicity of directed lights; the result is a confused visual effect.

3. Excess of diffuse lighting: depth perception is lost; the result is a very dull luminous environment (Figure 39).

Illuminance and evaluation of a lighting system

4. No shadows: a result of excessive diffusion, it flattens objects and hides surface details (Figure 39).

Figure 38: Still life lit by point-like source. Notice the shadow and the highlighting of any irregularity in the surfaces. It is also quite evident that the plate is chipped (look at the yellow circle).

> `Modelling` is defined as the ratio of cylindrical to horizontal illuminance at a point. (28)

Good modelling has values between 0,30 and 0,60 for a uniform arrangement of luminaires or roof lights.

Small values mean that most of the light is directed vertically. This has multiple consequences: there is a shadow right under the object, the horizontal surfaces are well lit and any protuberance of the object will cast a shadow and be highlighted.

Values close to 1 mean that there is only diffuse lighting, so there will be no shadows, all surface details will be flattened, and depth will be harder to perceive.

Figure 39: Still life lit by diffused light.
Notice the absence of shadows, and the worsening of depth-perception. It is harder to notice that the plate is chipped.

Larger values mean that most of the light will be directed horizontally, so the horizontal surfaces will be dimmer than the rest. Beyond this, not much can be said without first measuring vertical illuminance in at least four directions at 90° from each other.

If the values measured are similar, there will be no shadows. This is a good system to light a tall and cylindrical statue that must be surrounded by the viewers.

If there is a large difference, there will be a shadow in the direction of the smallest value, and a main direction of viewing the object from the direction of the highest value to the direction of the lowest one. This is a good system for a large and again cylindrical statue, only this time placed next to a wall, so that there is a main viewing direction.

If the task is not cylindrical, such as a statue with its arms spread out, the lack of vertical illuminance will show, and this levels of modelling will not be appropriate.

Illuminance and evaluation of a lighting system   73

Something similar happens with surface detail and depth perception: where there are large differences in vertical illuminance, depth perception will be easier and surface roughness highlighted.

6.12 Isolux curves

We need a way to see the illuminance on the task surface to evaluate the performance of the lighting system. We could use an illuminance table, and read and compare all the values printed in a Cartesian system onto the surface (see Figure 40).

0.484	120	148	177	224	348	474	601	731	874	1043	1213	1380	1527	1475
0.453	118	146	175	216	339	463	589	715	849	1011	1169	1310	1401	1375
0.422	115	143	171	200	319	441	563	685	802	951	1091	1204	1267	1250
0.391	111	138	166	194	292	409	526	642	752	872	992	1084	1131	1119
0.359	105	132	159	186	256	368	480	588	689	783	881	956	994	985
0.328	98	124	150	177	213	320	425	526	619	703	773	827	858	850
0.297	90	115	141	166	192	266	365	458	542	618	679	721	742	737
0.266	81	105	130	154	178	207	299	385	461	529	583	620	638	634
0.234	76	99	122	145	168	190	233	312	381	442	490	523	539	535
0.203	65	87	109	131	153	174	193	233	295	351	394	423	437	434
0.172	63	75	96	117	137	157	175	192	209	259	298	324	337	334
0.141	60	65	82	102	121	139	156	171	182	193	203	227	238	235
0.109	57	62	68	86	104	121	137	151	160	170	178	183	187	186
0.078	54	59	63	70	87	103	118	131	139	148	155	160	163	163
0.047	51	55	59	63	70	85	99	111	118	126	133	137	140	140
0.016	48	52	56	59	63	67	79	91	97	105	111	115	118	117
m	0.020	0.059	0.098	0.137	0.176	0.215	0.254	0.293	0.332	0.371	0.410	0.449	0.488	0.527

Attention: The coordinates refer to the image above. Values in Lux.

Grid: 128 x 128 Points

Figure 40: Illuminance table for a system.

Considering that Figure 40 is less than a twentieth of the total table, it is clear that this is a uselessly complex way of looking at things. The solution is to try and go for a different representation: the Isolux curves.

> `Isolux curves` are lines on the surface considered that join points sharing the same illuminance, usually indicated in the diagram itself. (29)

This new representation (Figure 41) contains more information than the table and it is much easier to read.

Figure 41: Isolux curve for the same system in Figure 40.
Notice the bad uniformity.

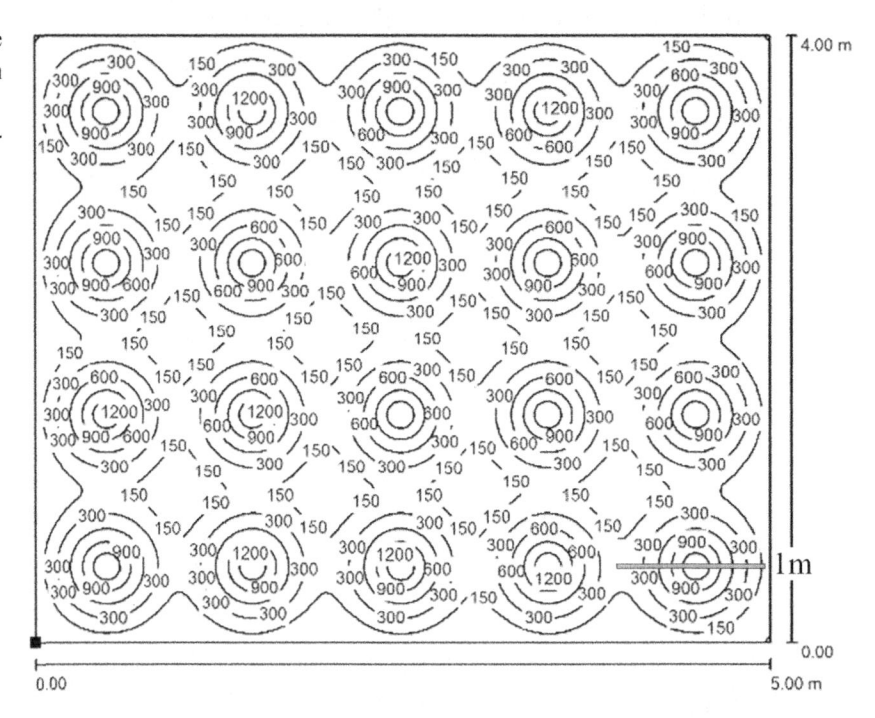

Let's detail the kind of information that we can gleam from the isolux curves:

1. Exact position in the surface of a particular level of illuminance.
2. Rate of change of illuminance on the surface.
3. Range of values for the illuminance in a certain spot.

Let's look into each.

6.12.1 Position of a particular level of illuminance

This is the actual definition of isolux: the locus of the points sharing the same illuminance. It's just a matter of looking at the legend.

The regulations (EN 12464-1, 2011) require that illuminance on major surfaces always stays over 50 lux on the walls, and 30 lux on the

Illuminance and evaluation of a lighting system 75

ceiling. This is something we can immediately verify, by looking at the 30 or 50 lux isolux, if it is present in the graph.

The same regulations link a minimum level of average maintained illuminance to each activity. Obviously isolux curves do not indicate averages, but only the point values. Still if we see that a large part of the task is under the requirements, we may suspect that the surface is non compliant, at least because of lack of uniformity, and that we need to intervene.

6.12.2 Rate of change of illuminance on the surface

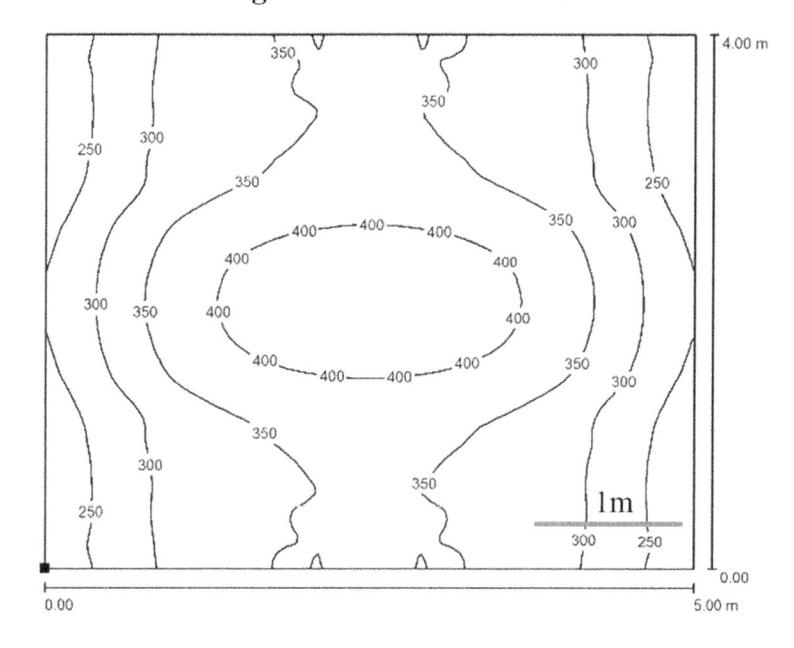

Figure 42: Isolux curves for a system with good uniformity.
Notice how the values change slowly as one moves around the interior.

Isolux curves contain indications about uniformity: if we see curves with very different values on their labels close to one another, we can deduce that illuminance has increased significantly in a small distance, i.e. that uniformity is bad; this may cause issues to visual comfort.

In Figure 41, in just a metre, the illuminance goes from under 150 lx to over 1500 then back to under 150 lx. This system will have very poor uniformity, in fact DIALux calculates an U_0 of 0.134

If on the other hand the values don't change so much, and the curves are not close to each other, we can conclude that the illuminance will remain pretty stable on the surface.

In Figure 42, in the same metre the illuminance goes from under 250 lx to between 300 and 350 lx. This means that the uniformity here will be rather good, in fact the simulation shows U_0 at 0.643.

6.12.3 Range of values for the illuminance in a certain spot

The illuminance in a certain spot will be between those of the two isolux curves closest to it that border the region where the point is. We can expect the actual value to be closer to the one of the nearest curve, but this is not certain.

6.13 Self evaluation

Answer the following questions to test your understanding of the material:

1. Describe the concept of visual task, making examples of surfaces that can be or not be one according to the destination of use of the interior or any other factor that comes into your mind.
2. Define illuminance and explain its meaning.
3. Provide examples of situations and activities requiring different levels of illuminance.
4. Describe the different types of illuminance, their uses and provide multiple examples for each.
5. Define the maintenance factor and explain what it is for.
6. Discuss how to evaluate a lighting system using uniformity.
7. Define modelling and discuss its effects.
8. Explain what an isolux curve is and what kind of informations one can obtain from it.

Intensity and illuminance

Let us consider the relation between the light source and the task and see where common sense takes us.

Suppose you are in a dark room at night, with just a candle as light source, and you want to read a book. In order to read the text on the page, you must move very close to the candle, and orient your book so that the light illuminates the page.

Let's try and translate what happened into the new illuminating language: the candle emits a small amount of luminous flux hence the intensities will be relatively small. In order to increase the illuminance on the page to a level that is high enough for us to read, we need to:

1. Reduce the distance from the source: illuminance is dependent on distance.
2. Orient the page towards the candle: illuminance is dependent on orientation.

Let us try and detail those correlations in a more formal way.

7.1 Illuminance under a luminaire as a function of distance and luminous intensity

Let's imagine a spacious interior, so large that the reflection of light on the walls can be left out (the same thing happens with very dark walls and ceilings), a single light source and a task. The source is over the centre of the task, hanging from the ceiling at such a distance that again reflections can be left out.

The light intensity distribution of the source is known. The objective of the exercise is to find the illuminance in the centre of the task.

The relation is shown in Figure 43 and the formula in (30): Illuminance in P is the quotient of intensity at ɣ = 0° by the square of the distance between the task and the photometric centre O of the fixture.

$$E = \frac{I}{h^2} \qquad (30)$$

Figure 43: Illuminance and intensity just under the fittings.

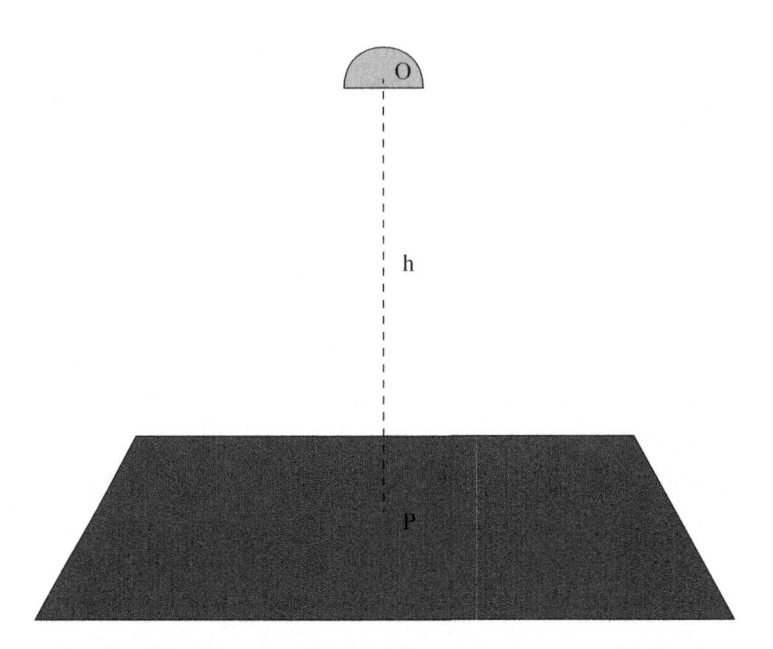

It is an inverse square law This formula means that illuminance changes with the square of the distance from the source, and linearly with the intensity. So if the distance doubles the illuminance will become a quarter, and if the intensity doubles, the illuminance will double as well.

Let's use this relation: the distance between the fixture and the task is two metres. We select the spotlight in Figure 44, its lamp emits 4000 lm, and we measure 241 cd/klm in the direction γ = 0° (point 1). What is the illuminance in the centre of the task?

Intensity and illuminance

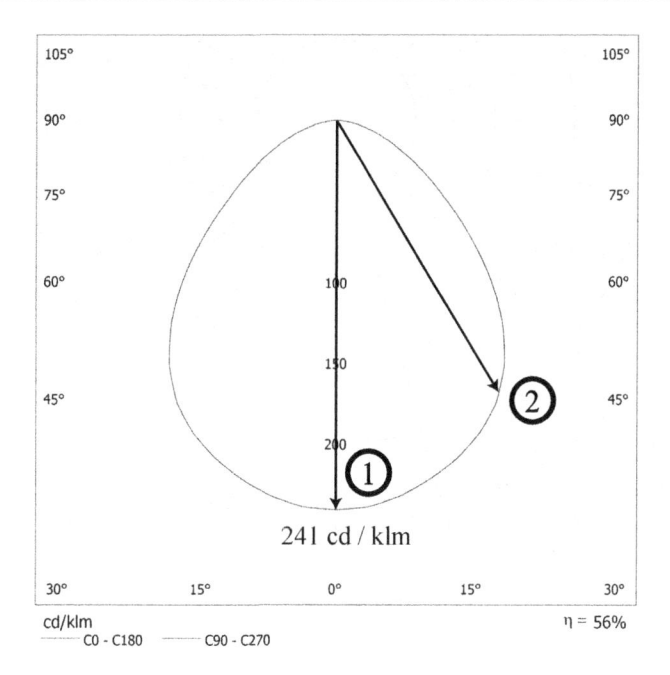

Figure 44: Light intensity distribution curve of a wide beam spotlight.
1: Intensity under the fitting
2: Intensity at 30°

Using the calculations in (31) and (32) the illuminance turns out to be 241 lx.

$$I = \Phi \; I_{spec} =$$
$$= 4000 lm \; 241 cd/klm = \quad (31)$$
$$= 964 cd$$

$$E = \frac{I}{h^2} =$$
$$= \frac{964 cd}{(2m)^2} = \quad (32)$$
$$= 241 lx$$

We might find ourselves in a different situation: let's say we wanted to choose the luminaire that produces 500 lx in the centre of the task. We need to invert (30), and we obtain (33).

$$I = E\ h^2 \tag{33}$$

Let's substitute illuminance and distance into the formula in (33).

$$I = E\ h^2 = 500 lx\ (2m)^2 = \\ = 2000 cd \tag{34}$$

We obtain 2000 cd of luminous intensity. It is about eight times as much as what the spotlight in Figure 44 produces. This fitting has a 35 W metal halide lamp, and we will learn, later on, that this is a very efficient lamp. To get eight time as much flux, we would need a 280 W lamp, but there is no such thing.

The solution is to change luminaire. Select a fitting with a narrower beam that still fits the task, because it will be able to concentrate the luminous flux in a smaller cone, thus providing higher illuminance. Failing that, use multiple luminaires or a different light source.

7.2 Illuminance and intensity: general formula

What if we need to calculate illuminance on a point off to the side?

Let's see (Figure 45) what happens when angle θ grows:

1. The distance between the fixture and the task increases. We already know from (30) that this means a reduction of illuminance given the intensity.
2. The task does not receive light face-on anymore (see Figure 46). The source moves closer and closer to the horizon of the task. This means that the same luminous flux will be spread over a larger surface, thus reducing illuminance.

This means that if the fixture emits constant intensity for any θ angle (see Figure 45), the illuminance will be highest right under the lumi-

Intensity and illuminance

naire, where the distance is smallest and the task is perpendicular to the source. Illuminance will then decrease as θ grows.

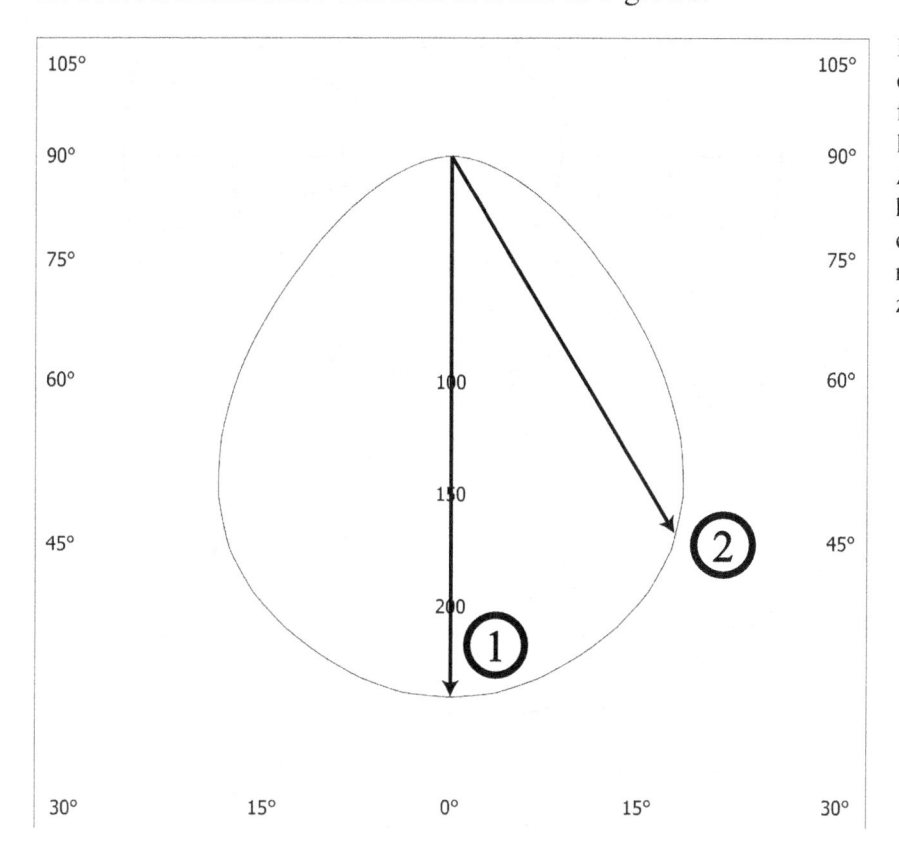

Figure 45: Illuminance on a point off to the side from the nadir of the light fitting.
As θ grows two things happen: the distance increases and the source moves towards the horizon of P.

Formally, the formula is:

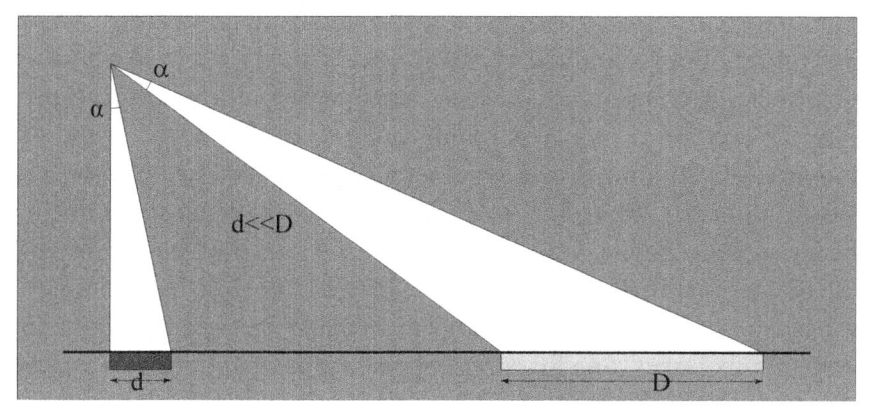

Figure 46: Two cones of light with the same vertex angle α emit the same luminous flux. A surface less perpendicular will spread the flux on a larger area (D) than a more perpendicular surface (d), so the illuminance in D will be smaller than in d.

$$E = \frac{I}{h^2}\cos^3\theta \tag{35}$$

Let's apply the formula to a continuation of the example in the previous paragraph, and put θ at 30°. Measuring Figure 44 at γ = 30°, we get an intensity over flux of 200 cd/klm (point 2).

In (36) the unit conversion from klm to lm is detailed in grey. Illuminance is 130 lx.

$$I = 200\frac{cd}{klm}\ 4000lm\ 0.001\frac{klm}{lm} = \tag{36}$$

$$E = \frac{I}{h^2}\cos^3\theta = \frac{800cd}{(2m)^2}\cos^3 30° \approx 130lx \tag{37}$$

The inversion of (35) is left as an exercise to the reader[1].

7.3 Cone diagrams

The cone diagram (Figure 47) is a simple way to display the behaviour of a luminaire, usually a spotlight, and it makes selecting the right fixture for the task much easier.

The graphical look of the diagram can change significantly, while the information contained is always the same. I will refer to the diagrams DIALux makes, such as the one in Figure 47. From the left, the column A indicates the distance from the centre of the luminaire; the column B is the diameter of the usable bright spot the fixture lights on the task. The grey cone shows the aperture of the beam spread, the exact angle (24.2°) is in C on the bottom row of the diagram, called half-value angle. Please note that this is not half the beam spread, it is the actual beam spread. The rightmost column contains six boxes, one for each distance value of column A. In each

---

1. For the answer see (38) on page 86.

Intensity and illuminance

box there is the illuminance in the centre of the task (D), the average illuminance in the usable bright spot (E), and half the beam spread (F).

Distance [m]	Cone Diameter [m]		Illuminance [lx]	
A 0.5	B 0.21	D E(0°) E(C0)	12.1°	45333 21254
1.0	0.43	E(0°) E(C0)	12.1°	11333 5313
1.5	0.64	E(0°) E(C0)	12.1° F	5037 2362
2.0	0.86	E(0°) E(C0)	12.1°	2833 1328
2.5	1.07	E(0°) E(C0)	12.1°	1813 850
3.0	1.29	E(0°) E(C0)	12.1°	1259 590

——— C0 - C180 (Half-value Angle: 24.2°) C

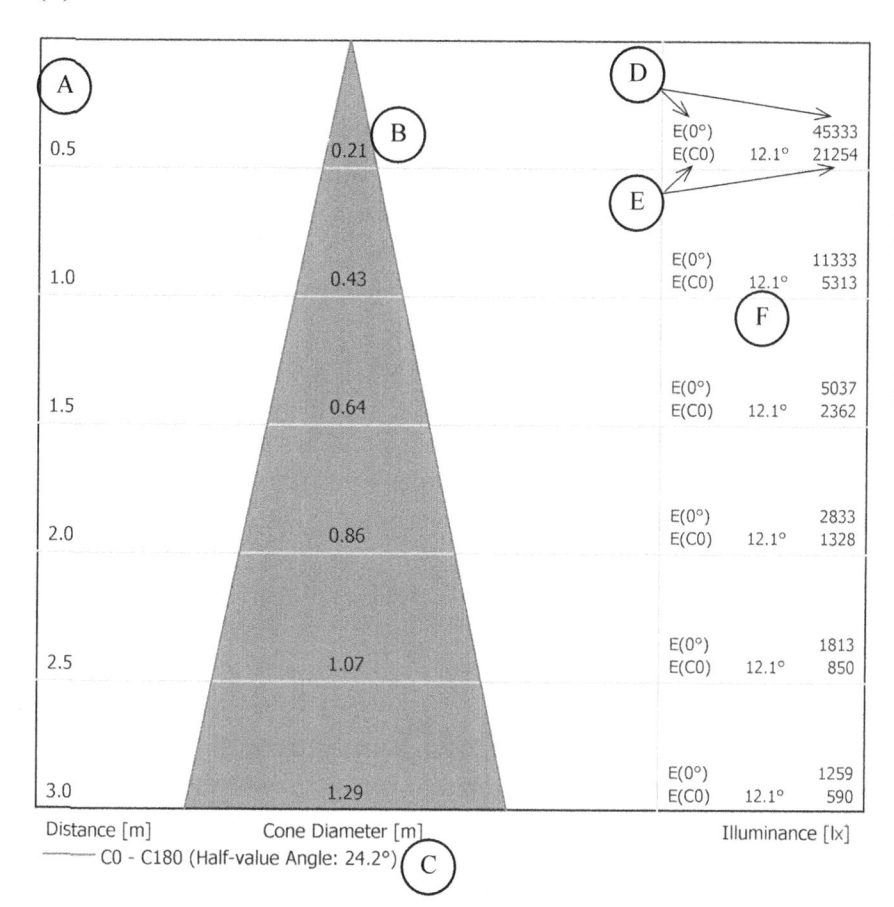

Figure 47: An example of cone diagram for a narrow beam spotlight. It contains the following: distance from the fitting (A), diameter of the bright spot (B), beam spread (C), illuminance in the centre (D) and average illuminance (E) in the pool of light, half beam spread (F).

Recalling (30), notice that as the distance doubles the illuminance becomes a quarter. The explanation for this is that the amount of luminous flux in the cone is constant, but the surface over which it is spread becomes four times bigger as the distance doubles.

This diagram shows the ability of the spotlight to provide a certain illuminance (on average or in the centre) at a certain distance, illuminating a certain diameter. In order to decide whether the spotlight is fit for the task, the designer can select any of these data as starting point and obtain the others quickly.

7.4 Illuminance uniformity and light intensity distribution

Illuminance uniformity is a very important objective for good, efficient, and functional lighting and it becomes an important criterion for selecting the luminaire. Let's see how.

As always let's start by analysing the situation according to our knowledge and common sense. We plan a horizontal task with a luminaire placed at a distance h, just like in Figure 45. We already know that all points with θ greater than zero will be farther away from the luminaire as θ grows, and that they will see the luminaire closer to their horizon, so the luminous flux of the fixture will be spread out on a larger surface (Figure 46). This means that we want a fixture that emits higher and higher intensities as θ grows to counterbalance these two effects. A deeper look into the formulas can detail this variation. We start by inverting (35) and finding (38).

$$I = \frac{E}{\cos^3 \theta} h^2 \qquad (38)$$

Since both the illuminance and the distance are constant, we can say that intensity must be proportional to the inverse of the cube of the cosine of angle θ. We need to plot this function to see how it looks and to be able to compare it to the light intensity distributions for the fixtures. The result (for θ smaller than 50°) is shown in Figure 48.

A memory trick to remember the shape of the curve is that it must look like the archetypal villain's moustache…

When the task is not aligned to the opening of the luminaire, the shape of the ideal light distribution curve remains the one in Figure 48, it only tilts by an angle so that γ = 0 is aligned to the normal of the surface of the task.

Intensity and illuminance

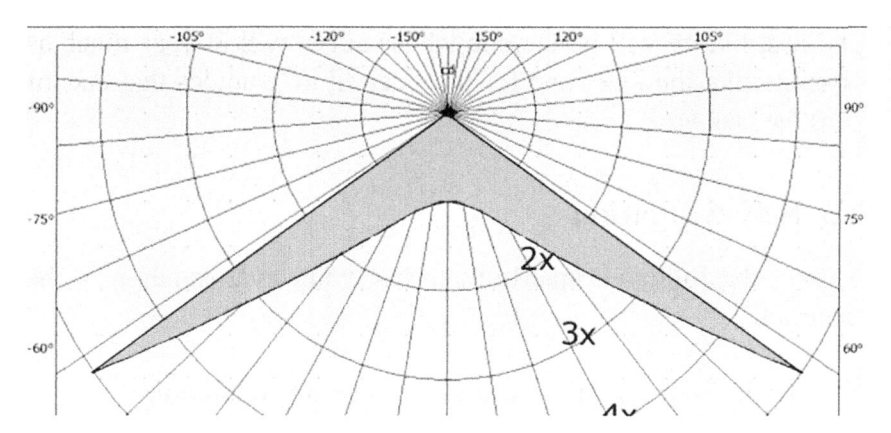

Figure 48: Shape of the ideal photometric curve for a horizontal task.

A particularly interesting case is when the fixture is ceiling mounted and the task is parallel to a wall, a type of fixture we will learn to recognise as a wall-washer later in the book: the curve will be tilted 90° clockwise and the half that would go over the ceiling cut out, so that it will look like Figure 49.

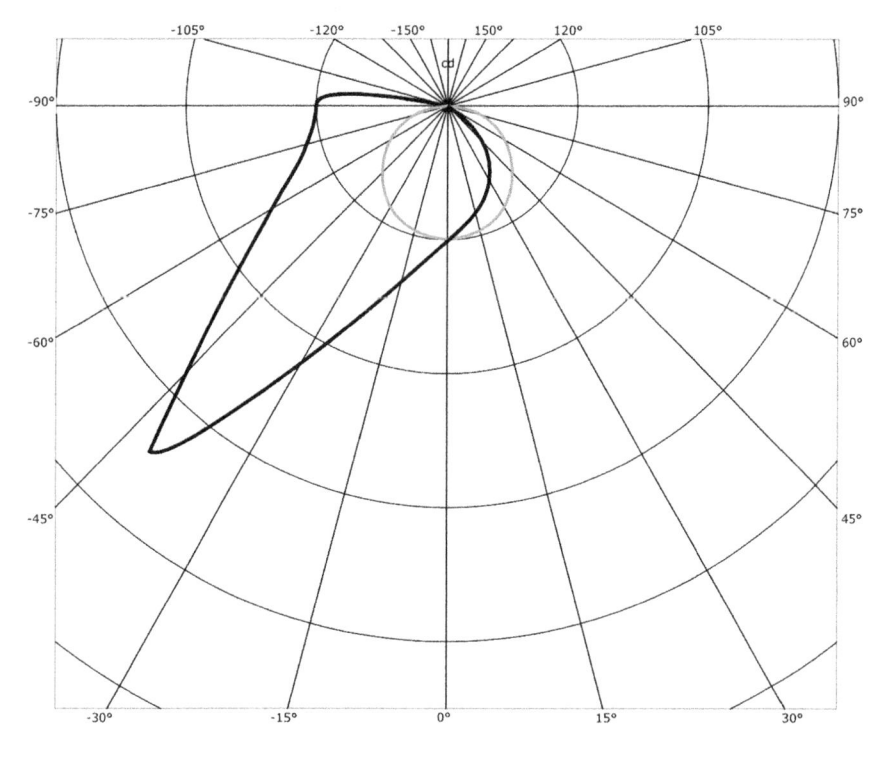

Figure 49: Light intensity distribution curve for a wall-washer light fitting.

The best fixture will be the one whose curve will look as much as possible like the ideal one for the interval of γ angles that encompass the task area.

7.5 Self evaluation

Answer the following questions to test your understanding of the material:

1. Explain how illuminance and intensity are connected.
2. Describe the relationship between intensity and illuminance right under the luminaire.
3. Describe the general relation between illuminance and intensity.
4. What is a cone diagram? What information does it provide?
5. Describe how the light intensity distribution curve should look to give illuminance uniformity, explain the link and give examples.

Elements of reflection

We need to introduce another performance indicator to evaluate visual comfort. A significant percentage of the field of view is occupied by the multiple surfaces of the interior (walls, windows, curtains, doors, furniture etc.). These surfaces will change the light, so knowing their properties is very important for the interior designer.

I will try to clarify two points:

1. How much light is reflected by a surface
2. Where the reflected light will go.

8.1 Reflectance

To find out how much light is reflected from a non light emitting material, we need to measure two quantities: the luminous flux that falls onto the material, and the luminous flux emitted by it. These values define the reflectance of the material.

> **Reflectance** (ρ) is the ratio of the reflected luminous flux to the incident flux.
> Reflectance is measured either as a percentage or as a number with $0 < \rho < 1$. (39)

We can roughly divide materials in reflectors and absorbers according to their reflectance.

> **Reflectors** are materials with $\rho \geq 0.5$
> **Absorbers** are materials with $\rho < 0.5$ (40)

Table 9 details the reflectance of common materials.

Table 9: Example of reflectance of a few materials.

Material	Reflectance (%)
White wall paint	85
Red bricks	30
Carrara marble	75
Dark polished wood	7
Light wood	40

Regulations[1] suggest average values for reflectance when doing back of the envelope calculations for interiors:

1. Ceilings from 0.7 to 0.9
2. Walls from 0.5 to 0.8
3. Floors from 0.2 to 0.4

When using a light calculation software, choosing the correct reflectance helps significantly in obtaining a realistic result.

8.2 Types of reflection

A polished metal surface reflects light in a different way than a plaster wall. Let's introduce three modes of reflection.

8.2.1 Specular reflection

Ideal specular reflection happens when the reflected ray of light has the same angle as the incident one. This happens obviously with mirrors and polished surfaces in general.

The main property of this type of reflection is image forming: an image of the source forms on the mirror, so that brightness changes significantly with direction. If I move even slightly away from angle α in Figure 50 I will not see the reflected ray anymore.

---

1. (EN12464-1 2011)

Elements of reflection 93

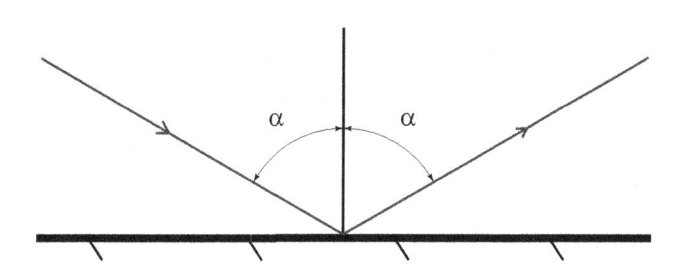

Figure 50: Specular reflection: the angle of incidence is equal to the angle of reflection.
This happens with polished surfaces, and it allows the formation of an image.

The shape of the surface determines how light will be reflected, with a very high sensitivity: a minor difference in the shape will significantly change the emission of light. Some reflectors are made according to this principle.

8.2.2 Diffuse reflection

Ideal diffuse reflection (Figure 51) happens when a single ray is scattered in all directions by the surface, with its intensity proportional to the cosine of the normal angle, so that the maximum intensity of the reflected scatter will always be perpendicular to the surface, and it will decrease and reach zero as we move from the normal to the tangent of the surface.

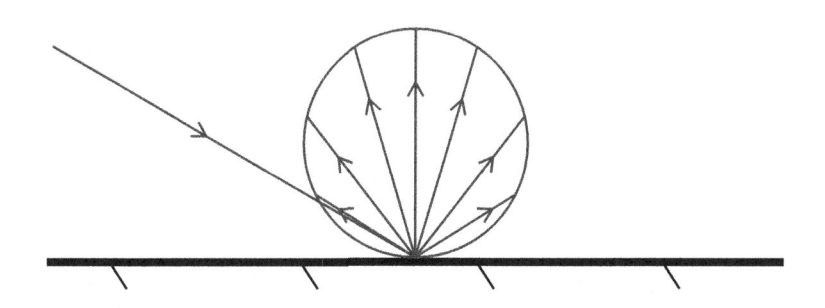

Figure 51: Lambertian reflection: a ray of light is diffused in a peculiar way.

Alabaster is a material that approximates this kind of reflection.

8.2.3 Mixed reflection

Obviously it is a mixture of the two previous types: there is no image formed on the surface, but the intensities are spread around a maximum direction (Figure 52).

It is the most common by far.

Figure 52: Mixed reflection: it is the most common.
There is diffusion but there is also a main direction of reflection.

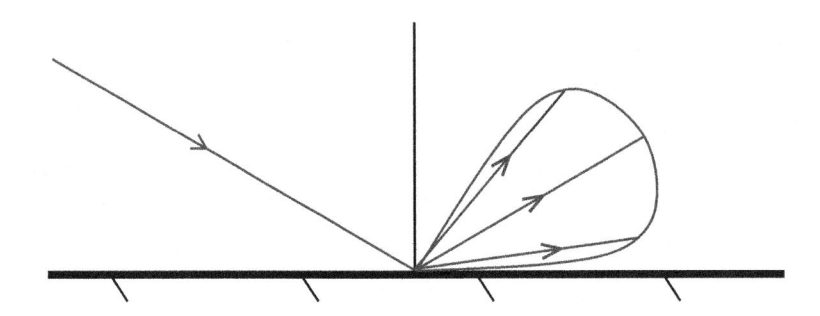

The shape of the surface determines almost completely the way in which the light is reflected.

A sheet of metal painted matte white will behave like this.

8.3 Reflection diagram

Reflectance is mostly due to the type of material chosen, while the type of reflection has more to do with the surface finishing of the object.

Elmer[1] observes that nearly perfect specularity can be found in both a nearly perfect reflector (silver at 98% reflectance), and in a nearly perfect absorber (black marble at grazing incidence). Similarly, near perfect diffusion can be found in both a near perfect reflector (magnesium oxide, used by gymnasts and climbers to improve their grip), and a near complete absorber (carbon black).

---

1. (ELMER 1980)

The combinations are shown in Table 10.

	Reflector	Absorber
Perfect		
Specularity	Silver	Black marble
Diffusion	Magnesium oxide	Carbon black

Table 10: Independence between reflectance and mode of reflection.

8.4 Self evaluation

Answer the following questions to test your understanding of the material:

1. Explain what is reflectance and what affects it.
2. Explain what types of reflection are there, and give examples of each.
3. Describe whether the type of reflection and the amount of reflectance are linked or not, and give examples.

Luminance

9

We looked into luminous flux to be able to evaluate lamps, we studied intensity to learn about the different fixtures, and we analysed illuminance to understand the performance of the lighting system. We found out that by following prescriptions on illuminance, uniformity, and colour rendering we can be sure that the task activity will be feasible. We know that uniformity and modelling give us an idea of how comfortable it will be to actually perform the activities, but there still is something missing.

We haven't yet found a way to evaluate the impact of the materials, their reflectance and type of reflection, on the visual comfort of who uses the interior.

9.1 Luminance

We need to introduce a new indicator that allows us to measure the effect of certain decisions the designer might take. For example, let's suppose a designer wanted to maximise the apparent size of the interior, and placed mirrors on all the walls and interior partitions of an office; and, to make the most of daylighting, he put no curtains at all on a very large window facing west. Would the office be comfortable? And how do we find out for sure?

We might begin by examining the different effect sunlight has on an observer's eye if it is reflected by a mirror hanging from a plaster wall (total blindness) or by the plaster wall itself (a sensation of diffuse brightness with no damage to visual performance). The comfort of the observer is very different, even if both the mirror and the wall receive the same amount of luminous flux and have the same illuminance.

What changes between the mirror and the plaster wall is the amount of luminous flux going into a specific direction (intensity) and the apparent surface that is emitting light.

Let's look into this:

1. Direction: It is enough to move slightly from the direction of the reflected rays of the mirror to significantly reduce the intensity reaching the eye of the observers, and getting out of the blinding reflection; while walking around a plaster wall will not change the intensities much.
2. Emitting surface: the size of the sun on the mirror will be really small and extremely bright, while the size of the bright area on the plaster wall will be much larger.

Brightness will then depend on the size of the apparent surface that is emitting the light, and on the amount of intensity going in the direction of the observer.

Figure 53: Apparent surface.

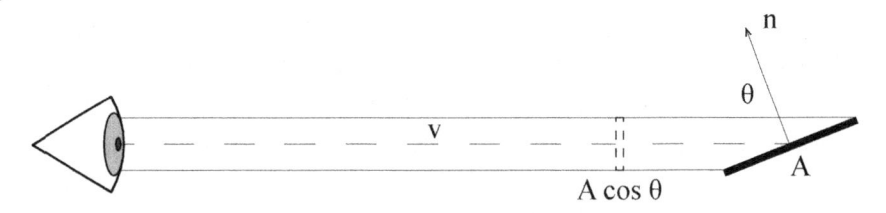

Let's start with the definition of apparent surface.

$$A' = A \cdot \cos\theta \tag{41}$$

In Figure 53: there is a surface (A) that emits light, it could be reflected from a luminaire, from the Sun etc., or the surface itself could be a light source; and the eye of an observer receives it.

There is an angle θ between the direction of observation (v) and the normal to the surface (n), so the apparent surface area that the observer sees is A cos θ.

Luminance

If the surface is perpendicular to the eye, then the apparent surface coincides with the actual surface (cos 0 = 1); if the surface is parallel to the direction v, then the observer will see just a line, and the area of the surface will be zero (cos 90° = 0)

We can now define[1] luminance in a given direction at a given point of a surface, with the following formula.

$$L = \frac{I}{A \cos \theta} \quad (42)$$

In (42) I is luminous intensity, A is the area emitting or reflecting light, and A cos θ is the apparent size of the surface as seen by the observer.

The verbal definition of luminance is given in (43).

```
Luminance is the quotient of the luminous
intensity in a specific direction by the
apparent bright area.                        (43)
It is measured in candle over square metre
(cd/m²).
```

Figure 54 shows definition (43): in grey the light intensity distribution of the surface, and the large arrow indicates the intensity emitted by the surface towards the observer (I_θ), the surface A is tilted towards the viewing direction v by an angle θ.

The luminance perceived by the observer will be directly proportional to the intensity in the direction θ, and inversely proportional to the area of the apparent surface A cos θ.

---

1. (EN12665 2011)

A candle over square metre is very small: the full moon reaches 2500 cd/m², an actual wax candle is around 20000 cd/m², and the sun is around 1.5 billion candle over square metre.

Luminance can be thought of in two ways depending on whether the concept is applied to objects or light sources. For light sources it describes the amount of visual contrast a fixture will produce: when luminance is very high one can expect a very directional light and harsh shadows, when luminance is low the opposite is true.

Figure 54: Intensity emitted by the surface in the direction of the observer.

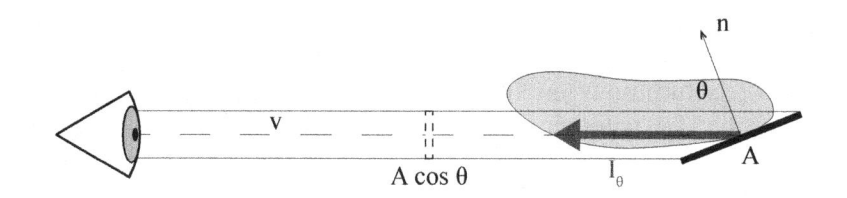

For non-light producing items, it describes how the illuminated object (be it a wall, curtain, task, etc.) modifies and emits back the light it receives. For example, a mirror-like surface will produce higher luminance and large variations of it when the direction changes, while a plaster diffuser will do the opposite.

It is interesting to note that a perfect diffuser (Lambertian in 8.2.2 on page 93) will have the same luminance no matter the direction of view. The proof is in (44).

$$\begin{aligned} L &= \frac{I}{A\cos\theta}; I = I_0 \cos\theta \\ L &= \frac{I_0 \cos\theta}{A\cos\theta} = \frac{I_0}{A} \end{aligned} \quad (44)$$

9.2 Luminance contrast

Suppose we lit a candle (20000 cd/m²), at night, under the moonlight (at most 2500 cd/m²): the glare would blind anyone looking at it, while the same candle, lit on a sunny day (at most 1.6 billion cd/

m²) would be almost invisible. This tells us that the absolute value of the luminance of the candle (20000 cd/m²) can be either blinding or unnoticeable depending on the background. This means that what counts is the variation of luminance, rather than its absolute value.

We can define luminance contrast[2] for a simultaneous viewing of two different levels of luminance in a visual field with (45).

$$C_2 = \frac{L_F - L_B}{L_B}$$

```
L_B is the luminance of the background, or    (45)
the largest part of the visual field
L_F is the luminance of the foreground.
3 ≤ C_2 ≤ 5 for best results.
```

The denominator of the definition accounts for the adaptation of the eye to the background level of luminance. The response of the eye to an increase in luminance depends on the absolute level at which the eye is used to: the higher the level, the larger the minimum difference in luminance must be to be perceivable.

The contrast will be larger when the difference between the foreground and the background, divided by the level in the background, will be large. Let's see in more detail the meaning of luminance contrast.

1. A luminance contrast close to zero means that there is very little difference between the foreground and the background: there will be no highlighting of the foreground.
2. When the luminance contrast is a large positive number, it means that the foreground will be much brighter than the background, e.g. when a narrow beam spotlight highlights a small task in a dark field of view.

---

2. (EN12665 2011)

3. When the contrast is a large negative number, it is the other way around: this time a very bright task has a dark hole in it.

4. Luminance is a major factor in defining visual acuity and adaptation. An increase in luminance requires the adaptation of the eye, a process that takes more time at lower luminance (up to thirty minutes). An increase in luminance corresponds to an increase in visual acuity, but the return grows smaller, up to an asymptotic maximum.

9.3 Self evaluation

Answer the following questions to test your understanding of the material:

1. Explain what is the apparent surface.

2. Explain the concept of luminance, and how it changes with intensity and angle between the surface and the observer.

3. Explain what is luminance contrast and what different values of it mean,

Visual Comfort

10

This chapter explains what is the comfort produced by lighting and how to design a comfortable system, and how to find out whether it is acceptable.

10.1 Visual comfort

Luminance is an objective way to measure visual comfort.

> ```
> Visual comfort is the subjective condition
> of visual well being induced by the envi-    (46)
> ronment.
> ```

Strictly speaking, it is a condition that varies with each observer.

Low visual comfort can impede the performance of any activity, just like lack of illuminance, so it is not just a matter of pure well-being, and it is not something the designer can do without and expect the interior to function. Visual comfort is not just about feeling well, it's also about performance.

While it is subjective, there are a series of pitfalls that should be avoided.

10.2 Veil

> ```
> Veils (or veiling reflections) are specu-
> lar reflections that appear on the object
> viewed, and that, partially or wholly, ob-   (47)
> scure the details by reducing contrast.
> ```

Veils are caused by non-harmonic distributions of luminance in the visual field.

Figure 55: Change in transparency in a shop window due to the veil effect.
Notice the disappearance of the veil under the awning

A veil like effect is what makes nearly transparent curtains effective for privacy in daytime.

Figure 56: Veil reflection indoor: the high luminance areas appear on the glass and reduce its transparency.

During the day the high luminance outside is reflected on them, this creates a veil of luminance so people outside just see the curtain and those inside still see the view, because the curtain is semi-transparent.

At night the situation is reversed: now the higher luminance is indoors rather than outdoors, so those inside see the curtains, and for those outside the curtains go back to being transparent.

Veiling reflections are common in shop windows (Figure 55), because the luminance outdoors is always much higher than in the store. In order to reduce this effect, shopkeepers employ retractable

Visual Comfort

awnings that have the added advantage of protecting the merchandise against direct sunlight. Another solutions is using glass with an anti-reflection coating, but this can be expensive with a large window.

Indoor veil (Figure 56) is usually caused by the often unavoidable interaction between glass partitions, cabinets or cases.

In order for glass to be really transparent, the "curtain at night" effect is necessary: higher luminance inside the case and lower luminance outside. If this does not happen, either because the glass pane reflects luminance from outside or for any other reason, the glass will stop being transparent and the veil will appear.

10.3 Shade

The definition of shade (48) is common sense translated into a more precise language.

> **Shades** are areas with very low luminance compared to the background. (48)

Figure 57: Unfortunate shade effect on a shop window display.

When areas with low luminance form, they have two effects: a direct and an indirect one. The direct effect is creating black holes in the field of view; the indirect one is the change in the object's meaning. The mannequin in Figure 57 looks like something out of an horror movie, because the black holes on the eyes modify the message we perceive: from "woman" to "monster".

An excessive directional component of the light, or the wrong placement of the fixtures will cause this effect.

10.4 Flicker

Again, the definition (49) is just a more precise form of the commonly held one.

> `Flicker` is the impression of unsteadiness of visual sensation induced by a light stimulus whose luminance or spectral distribution fluctuates with time. (49)

Rapid variations of luminance, much quicker than the adaptation time of the eye[1], of a certain amplitude and frequency, can produce the stroboscopic effect, that is the apparent change of motion and appearance of a moving object.

In a club, pulsating flashing lights that shine on the dancers make them appear still, even when they are moving. The eye behaves like a camera with a very short shutter time.

This is particularly dangerous in factories and laboratories, because it might, for example, make a power tool with an exposed blade appear to be turned off while it is actually on.

Flicker can also be a trigger for photosensitive epilepsy patients.

---

1. The adaptation time of the eye depends of the starting and ending luminance, and it is measured in minutes.

Electromagnetic ballasts for fluorescent lamps (13 on page 145) used to cause flickering at 50 Hz or 60 Hz (depending on your country of residence) because the fluorescent lamp produced a series of flashes synced to the frequency of the mains. Electronic ballasts solve this issue.

10.5 Glare

Glare is the most common cause of visual discomfort. Everyone, while driving at night, has been flashed by a driver that forgot to dip his full beam headlights. The luminance of the headlamp is so much higher than the background that the oncoming driver is temporarily blinded.

> **Glare** is the condition of vision in which there is a discomfort or a reduction in the ability to see details or objects, caused by an unsuitable distribution or range of luminance, or too extreme contrasts. (50)

Two types of glare are recognised by the regulations:

1. Disability glare: it impairs the vision of objects without necessarily causing discomfort
2. Discomfort glare: it causes discomfort without necessarily impairing the vision of objects.

10.5.1 Disability glare

Disability glare is the most common, it is the one we experience with high-beam headlights at night, or in Figure 49, where the luminance contrast is not so large that it causes discomfort, but it is enough to completely erase any detail of the book.

This type of glare is due to light scattering inside the eye because of luminance contrasts. The most common causes are:

1. Spotlights: they could be

1.1. Wrongly positioned.

   1.2. Wrongly pointed.

   1.3. Wrongly chosen.

2. Not being fully aware of the relative positions of reflective surfaces (e.g. glass panes), the lighting system and the tasks;

3. Not having fully plotted the course available to the users of the interior and their fields of view.

Looking back to "8.2.1 Specular reflection" on page 92, it is easy to understand why surfaces with specular reflection are more prone to causing glare: the luminance is so high in a specific direction that excessive luminance contrasts can easily occur.

Figure 58: Disability glare on the case of a precious book: the high luminance on the glass makes it impossible to see anything of what is behind the glass.

10.5.2 Discomfort glare

Discomfort glare occurs when an extended region of the field of view is brighter than what we can normally adapt to, which is about

10000 cd/m^2[2]. For example, fresh snow in bright sunlight can reach a luminance of 30000 cd/m^2, or well beyond the comfort zone.

It is easy to identify this type of glare because it can be reduced or eliminated by a pair of sunglasses.

10.6 Bright lamps

Regulations impose guidelines to guarantee visual comfort in lighting installations, and solve specific problems that might arise.

A very high luminance light source, not properly screened, might cause light scattering in the eye and glare, even when not looked at directly.

Luminance of the lamp [kcd/m^2]	Minimum screening angle α [°]
20 to 50	15
50 to 500	20
Over 500	30

Table 11: Minimum screening angles for high luminance light sources, according to EN12464-1.

This is why the European standard (EN 12464-1 2011) requires that lamps have a higher screening angle as their luminance grows. The minimum screening angle, measured starting from the horizontal, is indicated in Table 11.

This is especially relevant when evaluating LED light sources. They do have low intensities, but they also have a microscopic emitting surface, which means very high luminance and a serious risk of glare. The damage to visual comfort is the reason why LEDs without diffusers must be out of sight. This is the most important issue with using LEDs in interiors.

---

2. (SMITH 2002)

Figure 59: High luminance lamp screening angle.

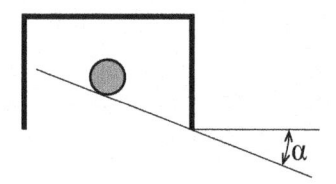

10.7 Luminance control for workstations with Display Screen Equipment

In these interiors, the operators will spend most of their time looking at a reflective computer screen on which both veil and glare can easily occur.

The causes for these failures in visual comfort are either light fittings that emit luminance in the direction of the screens, or reflections of artificial or natural luminance onto partitions, walls or ceilings placed so that the operator sees them on his display (Figure 60).

Figure 60: Glare on a display screen equipment.
On the left, the black luminaire emits rays of light at γ over 65°, which cause glare, or veil. On the right, it is a specular-reflecting partition that causes the problem.

Looking at the possible reflections, it is apparent that the more dangerous γ angles are those closer to the horizontal (γ = ±90°).

Since the monitors are almost vertical, excessive luminance emitted at or over γ = ±65° might be reflected onto the screen and then back on the worker. To avoid this problem, regulations impose a limit on luminance emitted by fixtures at or over 65° where DSEs are in use.

Figure 61: A not very tidy desk with two large high-luminance spots on the top monitors.

Not only there is no way to work comfortably, but the number of errors increases significantly.

There is more: not all conditions of use, screens and programmes are equal (Table 12). Some applications have higher necessities concerning colour and details of the shown information, such as CAD or colour inspection, so they need stricter requirements in luminance control, to warrant a better performance.

Moreover, if the display has its high state luminance larger than 200 cd/m^2, or the programmes used have positive polarity (white background and dark text), they will be better at reducing the damaging luminance contrast from stray light, so the requirements of the regulation will be less strict.

It is important to make sure that the luminaire selected satisfies the regulations. Manufacturers are supposed to declare conformity, but if they don't, the designer can either refer to the luminance table detailing the values of luminance for each angle, or he can look at a luminance diagram, either supplied by the manufacturer or calculated by the lighting software (Figure 62).

Table 12: Luminance limits for a lighting system in an interior where there are Display Screen Equipment (DSEs).

	At gamma ≥ 65°	
High state display luminance	High	Low
	L > 200 cd/m²	L ≤ 200 cd/m²
Case A Positive contrast and normal requirements on colour and detail of the information shown	≤ 3000 cd/m²	≤ 1500 cd/m²
Case B Negative contrast and/or high requirements on colour and detail of the information shown	≤ 1500 cd/m²	≤ 1000 cd/m²

Figure 62: Luminance diagram.
The γ = 65° line is inside the grey circle indicating 1000 cd/m².

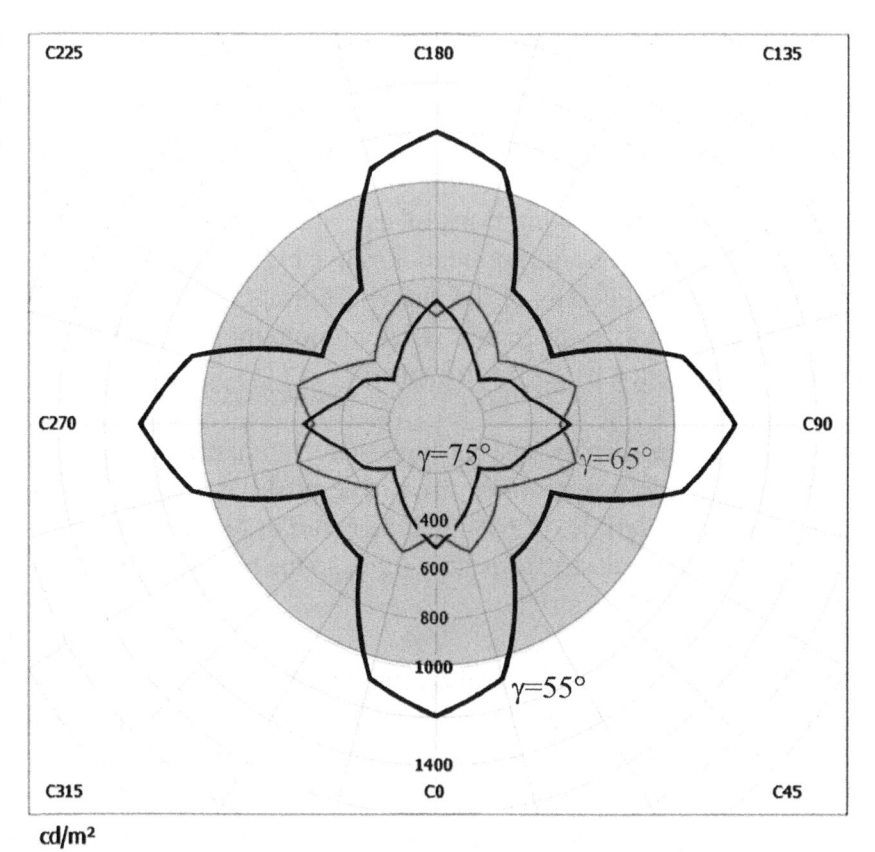

The consequences of these issues are crippling indeed (Figure 61).

Visual Comfort

Figure 62 shows the values of luminance for a γ angle of 55°, 65° and 75°. The line for γ = 65° (blue one) is inside the circle representing 600 cd/m², this means that the fixture is compliant even to the most restrictive case.

10.8 Offending zone

If we manufacture a mirror shaped like CIE's office task area (Figure 63), we can see on it the reflection of an area of ceiling immediately over the user. This is the offending zone (Figure 64): it is the area of the ceiling where, if we put luminaires there and if we had a visual task with any glossiness, the system would cause veiling reflections or glare on the task.

Figure 63: CIE office task area, with high luminance reflection of light fitting placed in the offending zone.

Traditionally the solution would be to use low luminance luminaires, such as fluorescents, or to avoid placing luminaires in the offending zone to minimise veiling reflections.

The problem with this approach is that to a certain extent veiling reflections give meaning to a scene. This high contrasts in the visual task (which is something we may like) comes from high contrast on the surrounding light field.

Figure 64: Offending zone for a person at a desk.

What is the solution, then? Once again the requirements provide it: where there is a need for high contrast, the designer may install high contrast luminaires in the offending zone and it will be the user's task to move his head and the task slightly to avoid the veiling reflections. In the rest of the installations, low luminance fittings will be best.

10.9 Unified Glare Rating (UGR)

In order to evaluate the visual well-being of an observer in an interior, complete with furniture, curtains and lighting system, there is a very practical aggregate indicator that we can use: the UGR value.

> **UGR** stands for Unified Glare Rating, and it shows the propensity of an interior to cause glare for a certain observer, with higher numbers showing more risk. (51)
>
> **UGR** goes from 10 (no glare) to 30 (pronounced physiological glare).

Visual Comfort

The UGR value is calculated by a formula that takes into account:

1. For the luminaires: their position relative to the observer and their luminance emission;
2. For the observer: his position and the direction of his line of sight
3. For the interior: its shape and the luminance of each surface.

It makes sense to expect that a critical activity will require an interior that has little propensity to cause visual discomfort, i.e. a low UGR value; while a simpler activity should be all right with less restrictive requirements.

The UGR is a very aggregate indicator: it will only tell you whether the interior works or not, it is unable to give any details about why it is not working[3].

In order to find out whether the interior is compliant, the designer must look up the maximum UGR accepted by the standards (e.g. the European regulations) for the activity that is going to be performed. For example a value no larger than 28 is mandated for a hallway, 19 for a conference room (where one is supposed to be doing more critical work than in the hallway) and 16 for a colour and textile inspection area.

We need to find out how one obtains the UGR value for an interior.

Manufacturers publish a UGR table for each luminaire (see Figure 65), and there is also a way to obtain it via the lighting calculation software. This table is obtained by presuming that a whole room will be illuminated by multiple rows of the fitting in question, and by giving coefficients that need to be applied in order to obtain a result for a specific interior. Unfortunately this manual method is quite cumbersome and it has limitations.

---

3. Paragraph "21.10 UGR evaluation" on page 488 will deal with improving the UGR for a room.

Table 13: Example of European regulation requirements for a retail premise.

Type of area, task or activity	$\bar{E}_m$ [lx]	UGR_L	U_0	R_a
Sales area	300	22	0,40	80
Till area	500	19	0,60	80
Wrapper table	500	19	0,60	80

The designer can also have the UGR values calculated for him by the lighting software, but it means that the interior must already be modelled in the program.

Figure 65: Example of UGR table for a luminaire.

Glare Evaluation According to UGR

ρ Ceiling		70	70	50	50	30	70	70	50	50	30
ρ Walls		50	30	50	30	30	50	30	50	30	30
ρ Floor		20	20	20	20	20	20	20	20	20	20
Room Size X	Y	\multicolumn{5}{c}{Viewing direction at right angles to lamp axis}						

Room Size X	Y	\multicolumn{5}{c}{Viewing direction at right angles to lamp axis}	\multicolumn{5}{c}{Viewing direction parallel to lamp axis}
2H	2H	21.3	22.4	21.6	22.6	22.8	21.4	22.4	21.7	22.6	22.8
	3H	21.2	22.1	21.5	22.3	22.6	21.3	22.2	21.6	22.4	22.7
	4H	21.1	22.0	21.5	22.2	22.5	21.2	22.0	21.5	22.3	22.6
	6H	21.1	21.8	21.4	22.1	22.4	21.2	21.9	21.5	22.2	22.5
	8H	21.0	21.8	21.4	22.1	22.4	21.1	21.9	21.5	22.2	22.5
	12H	21.0	21.7	21.4	22.0	22.3	21.1	21.8	21.5	22.1	22.5
4H	2H	21.3	22.1	21.6	22.4	22.7	21.3	22.1	21.6	22.4	22.6
	3H	21.2	21.8	21.5	22.2	22.5	21.1	21.8	21.5	22.2	22.5
	4H	21.1	21.7	21.5	22.0	22.4	21.1	21.7	21.5	22.0	22.4
	6H	21.0	21.5	21.4	21.9	22.3	21.1	21.6	21.5	22.0	22.3
	8H	21.0	21.5	21.4	21.8	22.3	21.1	21.5	21.5	21.9	22.3
	12H	21.0	21.4	21.4	21.8	22.2	21.0	21.5	21.5	21.9	22.3
8H	4H	21.0	21.5	21.4	21.8	22.2	21.0	21.5	21.4	21.9	22.3
	6H	20.9	21.3	21.4	21.7	22.2	21.0	21.3	21.4	21.8	22.2
	8H	20.9	21.2	21.3	21.6	22.1	20.9	21.3	21.4	21.7	22.2
	12H	20.8	21.1	21.3	21.6	22.1	20.9	21.2	21.4	21.7	22.2
12H	4H	20.9	21.4	21.4	21.8	22.2	21.0	21.4	21.4	21.8	22.2
	6H	20.9	21.2	21.3	21.6	22.1	20.9	21.3	21.4	21.7	22.2
	8H	20.8	21.1	21.3	21.6	22.1	20.9	21.2	21.4	21.7	22.2

Variation of the observer position for the luminaire distances S

S = 1.0H	+2.6 / -10.4	+2.0 / -4.1
S = 1.5H	+4.1 / -14.5	+4.2 / -10.1
S = 2.0H	+5.3 / -15.1	+4.7 / -11.1
Standard table	BK00	BK00
Correction Summand	2.8	2.9

Corrected Glare Indices referring to 9000lm Total Luminous Flux

I suggest a mixed approach: for a back of the envelope calculation, the designer should look at the values in the table and compare them directly to the requirement: if they are all smaller than the requirement, it means that probably the luminaire will be acceptable. If not it is better to proceed with the simulation and get precise values.

Finally we can understand all the regulation requirement for a certain activity (Table 13): Average maintained illuminance, maximum UGR value, illuminance uniformity and colour rendering index.

10.10 Self evaluation

Answer the following questions to test your understanding of the material:

1. Explain what is luminance contrast and describe how it can be an indicator of visual comfort.
2. Explain what is visual comfort.
3. Give examples of the four common pitfalls that should be avoided to maintain visual comfort in an interior.
4. Explain the difference between disabling glare and discomfort glare and provide examples.
5. Explain the reasoning behind luminance limitation for lamps.
6. Explain why and how luminance should be controlled in offices with display screen equipment, and provide examples.
7. What is the offending zone?
8. Explain what is the Unified Glare Rating and describe how it can be an indicator of visual comfort.

Part 2

Using the tools and concepts from part 1, I will describe the characteristics and performance of the types of lamps used in indoor lighting.

I will present a table of performance indicators, some already familiar to the reader, and some new, that makes it easy to select the most appropriate lamp for the task.

Figure 66: The incandescent lamp.

How is light produced and evaluated 11

11.1 Light production

The physical process used to produce light in a lamp will determine many characteristics of it: some can only output warm light, others generate a discontinued spectrum, some are inherently efficient, other remarkably less so.

Let's look into four ways of producing light.

11.1.1 Incandescence

We all know that heating a metal up makes it glow. Using a more precise language, it means that when an electric current goes through a conductor, this will produce heat according to a physical law (the Joule effect). What then usually happens is that the wire rapidly melts and the circuit is broken. If the conductor is made of a metal with a very high melting point, like tungsten, the circuit will not break so quickly and the lamp will continue producing light.

This process is used by incandescent and halogen lamps.

Because the colour temperature is, in this case, the actual temperature of the metal, and because tungsten melts at around 3400 °C (or about 3700 K[1]) halogens will never be able to produce cool light.

11.1.2 Fluorescence

There are some materials that emit light when they are hit by ultraviolet radiation. They are a combination of phosphorus and oth-

---

1. To get the temperature in kelvin, just add 273.15 to the temperature in celsius degrees.

er elements, and emit a different spectrum according to their exact chemical composition.

This phenomenon is well known, and the phosphors used to produce light have been perfected for a long time. This excellence in material engineering is what allows fluorescent lamps to be so flexible in the light they produce: they generate warm, neutral or cool light, high and low colour rendering, and special light for more particular applications.

This process is used on fluorescent lamps and some white light LEDs.

11.1.3 Electric arc

It is a very disruptive, explosive phenomenon that happens when the voltage between two electrodes is so high that the medium in which they are immersed (usually air or a mixture of noble gases in the case of a bulb) breaks into plasma, and a stable discharge appears. This produces tremendously high temperatures, and a prodigious amount of light. It is not unlike a tiny lightning bolt that does not end in a fraction of a second, but keeps going for as long as the lamp is on.

The gas in the burning chamber of the lamp determines the colour of the light. These lamps generate a discontinuous spectrum, but, because the mixture of gases can be tailored quite precisely to the application, the light produced is very high quality.

The electric arc is used in fluorescents and metal halide lamps.

11.1.4 Semiconductor physics

This last physical process has entered illuminating science recently, and it is the way in which LEDs produce light.

When two very special materials are mixed, the result, called a semiconductor, exhibits peculiar properties. The material develops holes[2] into which electrons may fall. When an electric current goes through the semiconductor, electrons can fall into holes and release a quantum of light, called a photon. Conceptually it is not too dissimilar from how Swiss cheese is a mixture of cheese and holes, and how holes can capture small peas rolling on the slice of cheese[3].

The characteristics of the light produced depend on the materials. Until very recently, this meant that the quality of LED light was rather average.

11.1.5 Types of lamps and processes

Table 14 summarizes what lamp uses what process.

Process	Type of lamp
Incandescence	Incandescent + Halogen
Electric Arc	Metal halide + Fluorescent
Fluorescence	Fluorescent + LED
Semiconductor Physics	LED

Table 14: Processes and types of lamps.

---

2. Actually lower energy states

3. If small peas glowed when they fell...

11.2 Evaluation of a lamp

The outline of my evaluation method for a lamp is presented in Table 15. Let's see the meaning of the various lines.

11.2.1 The shape

11.2.1.1 The bulb

Lamps come in standard shapes, some of which are indelibly associated not just to the type of source, but to the concept of lamp itself: the silhouette of the incandescent bulb, for example, has become a symbol for all lamps (Figure 66).

Table 15: Evaluation table for lamps.

Performance	Effect
Shape: bulb	Use
Shape: cap	Use
Colour temperature	Ambiance
Colour rendering	Light quality
Luminous efficacy	Costs
Average life	Costs
Electrical devices	Costs, reliability

Each bulb type has an international designation based on ILCOS (International Lamp Coding System). Table 16 shows a few examples of the most common appearances.

11.2.1.2 The cap or base

There are two major groups of lamps: naked (i.e. without integrated reflector), and with an integrated reflector (the one in the second row and second column in Table 16). Normally the latter are used in cheaper, less performing luminaires, because the reflector they contain is usually of much lower quality than the ones in luminaires.

How is light produced and evaluated 127

The cap (or base) is standardised as well, to allow easy replacement of failed lamps.

Table 16 shows a few examples of caps and their codes.

11.2.2 Correlated Colour Temperature

Correlated colour temperature is sort of a qualitative indicator of the closeness of the light to the cooler colours (2.6 on page 24). It is one of the first decisions one must take when developing a lighting system. At this time we consider the decision taken, we are only interested in finding out what CCT a specific source can emit, or, vice versa, what source do we need if we require a certain CCT.

Lamp	Bulb	Cap	Lamp	Bulb	Cap
	IAA	E27		FD	G5
	IAB	E14		HR-1	GU5.3
	HS	G9		MT	G12
	HD	R7s		FSD	2G11

Table 16: Examples of lamps with ILCOS code for bulb and cap.

11.2.3 Colour rendering

Colour rendering (2.5 on page 22) is only partially determined by the ambiance of the interior that the designer wants to create because each activity requires a certain minimum CRI: for a typical interior we can expect it to be around 80%.

Please remember that a low CRI lamp is a hard choice to defend today.

11.2.4 Luminous Efficacy

Luminous efficacy (4.2 on page 38) is a quantitative indicator that has risen to primary importance in the last few decades - it is now common knowledge to talk about energy saving lamps and LEDs.

In the last few years the lack of luminous efficacy has determined the phasing out of all the traditional incandescent lamps, and the replacement of a large percentage of halogens with more efficient versions. It is still an on-going process.

In my opinion it is interesting to see how the higher efficiency lamps compare in terms of total costs with the less efficient solutions, and how that comparison has changed in recent years. We shall look into this shortly.

11.2.5 Average rated life (B_{50})

It is the first of the new concepts I have to introduce to complete the table.

Average rated life is a statistical indicator of the lamp's duration: one takes a large number of sources and subjects them to as many on-off cycles, each six hours long, as needed. The reason for the six hours cycles is that the turning on and off wears out a lamp much more than staying on indefinitely, and it is a more realistic simulation of the life of the source. After a certain number of cycles, some lamps will have failed.

How is light produced and evaluated 129

> **Average rated life** is designated B_p, where P is a percentage, and it indicates the number of hours after which P% of the sources has failed.
> It is equivalent to the time after which a sample lamp in a very large population will have a probability of failure of P%

(52)

So, for example B_{50}, indicates the number of hours after which that 50% of the lamps will have failed. Figure 67 shows an example for a fluorescent lamp.

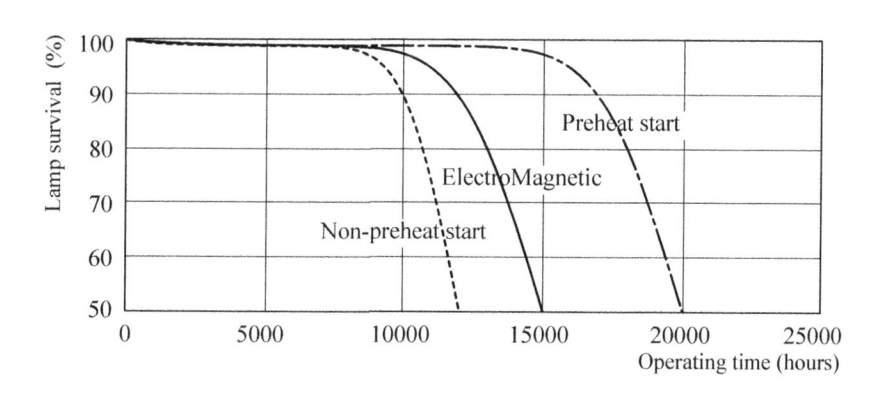

Figure 67: Different lamp survivals (B) for a Philips Master TL-D90 fluorescent lamp according to the type of ballast. Here B_{50} varies from 12000 to 20000 hours.

Together with average rated life another concept needs introducing. It is called rated lumen maintenance life, and it takes into account the fact that lamps emit less and less luminous flux as they grow old (Figure 68).

> **Rated lumen maintenance life** is indicated by L_Q, where Q is a percentage, and it is defined as "the elapsed operating time, in hours, at which the Q percentage of the lumen depreciation or lumen maintenance is reached."

(53)

So L_Q is the time after which the lamp produces Q% of its initial luminous flux.

The maintenance factor (see "6.8 Maintenance factor" on page 66), among other things, takes this into account.

Because the average rated life is such a simple and understandable indicator, it is sometimes misused both in meaning and in values. Concerning its meaning, I have seen it used as the only indicator of a good lamp. No mention was made of colour rendering, colour temperature, luminous efficacy, size, starting time and so on: the message was that the best lamp is the one that lives longer. As always, the wise designer should mistrust any decision based only on one indicator, because it is usually not the most informed one.

Figure 68: Rated lumen maintenance life (L) for a Philips Master TL-D90 fluorescent lamp. L_{90} is after about 20000 hours.

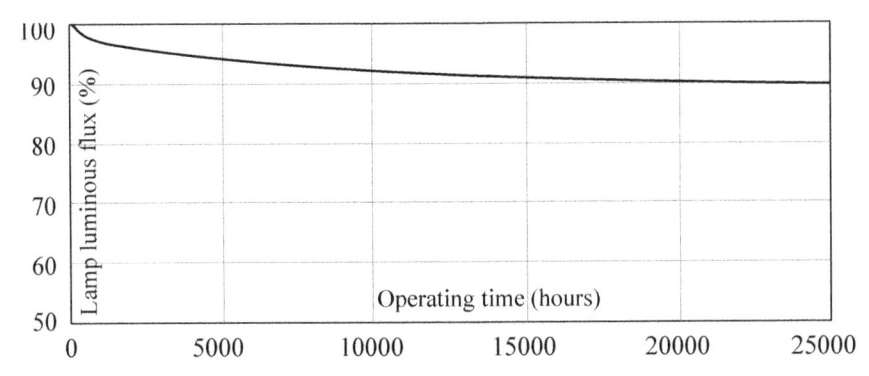

Concerning the values, LED technology are the culprits of much misuse. When they were introduced in lighting, the manufacturers declared an average rated life of 100000 hours. This number is enormous: it was more than five times longer than the previous longest lasting lamp. If one were to use such a lamp for eight hours each day, it would take more than 34 years before the source reached its average rated life. The numbers today have changed, and become a more sensible 30000 hours, and they will probably change again in the future. It is still more than one and a half time as much as the previous longest living lamp, and it means that it takes more than ten years to reach the average rated life with eight hours of use per day.

Finally I wanted to look into the relation between average rated life and maintenance intervals (i.e. the number of hours after which a man is called to clean the luminaires, replace the lamps and so on). The link between the two is more structured that what may appear.

On one hand, a shorter lasting lamp surely means a shorter maintenance interval, because the lamps will fail. On the other hand, a longer average rated life might not mean a longer maintenance interval, because there may be other issues that require intervention (e.g. dirt accumulation in the luminaires, or problems with the driver or ballast).

Obviously the maintenance interval is much shorter than average rated life: it would not be sensible to wait until half of the lamps have failed before calling maintenance.

11.2.6 Electrical devices

Some types of lamp can be run directly off the mains, others require one or more intermediate devices. These can be:

1. Transformer: it changes the voltage of the electricity for the lamp, e.g. a 12V lamp needs it to connect to the mains.
2. Capacitor: it is necessary for power factor correction and required by the public utility.
3. Starter: it provides the necessary high voltage discharges that turn on a lamp. Its function is sometimes taken over by electronic ballasts.
4. Ballast: its function is to keep the lamp on once it has started, and to make it work at its best.

Usually it is either a starter, ballast and capacitor system, or a transformer and capacitor one.

The interior designer doesn't need to know the details of these devices. The only decision he has to make is whether to select a luminaire with a traditional (electromagnetic) ballast or an electronic one.

Electromagnetic ballasts have been in use for many decades; their advantages are that they tend to be more robust and less sensitive to humidity. Their disadvantages are many:

1. They usually take longer to bring a lamp to full power.
2. They require the starter.
3. They can't keep the lamp in its ideal working conditions.
4. They don't stop trying to turn on a failed lamp, this can burn out the ballast and cause a minor risk of fire.
5. They can cause flicker, as they work at the mains' frequency.

Electronic ballasts have none of the disadvantages, and should be employed whenever the humidity allows it.

11.3 Self evaluation

Answer the following questions to test your understanding of the material:

1. Explain what are the four process used to generate light, what kind of light is produced, and in what types of lamps they are used.
2. List the characteristics of a lamp that contribute to the evaluation table.
3. Describe the shape of lamps.
4. Explain what is average rated life.
5. Explain what is rated lumen maintenance life.
6. Describe the devices that must be connected to certain types of lamps to make them work.
7. Electromagnetic ballasts: advantages and disadvantages.

Figure 69: Examples of halogen lamps without reflector, drawn to scale. The lamp on the bottom left of the picture is a low tension one.

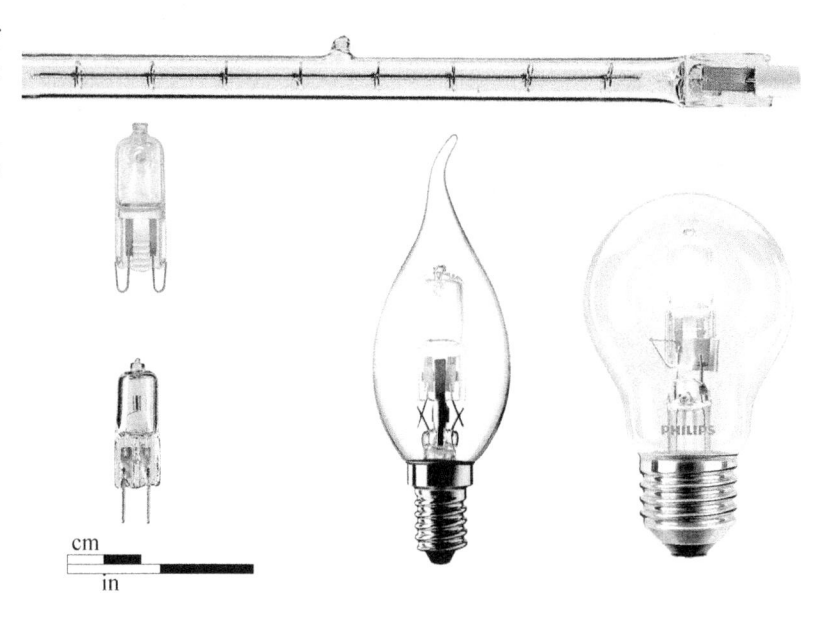

Figure 70: Examples of halogen lamps with reflector, drawn to scale. The three lamps on the left are low voltage, the others are mains.

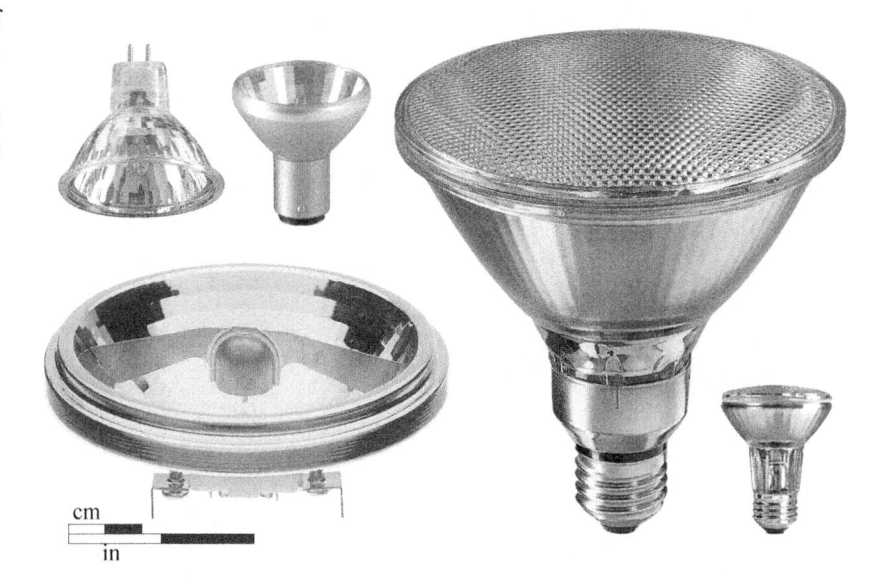

Halogen lamps

12.1 Notes on incandescent lamps

Before starting with halogens, I will give just a few notes on incandescent lamps. An international agreement has determined that they need to be retired because of their abysmally low luminous efficacy. The idea is to replace them with a more efficient type of lamp, such as the halogen (this will reveal its comedic value as you finish reading part 2), the compact fluorescent and the LED lamp.

The replacement can be difficult with decorative fixtures: incandescent lamps are easy to manufacture in artistic shapes, and when the shape is an integral part of the design of the luminaire, it may not be acceptable to make do with the simpler frosted bulb of a compact fluorescent (see Figure 76 on page 144), or the more complex double bulb of a halogen (see Figure 69).

A few special types of incadescent lamps can still be manufactured, such as the very low power versions that usually light home fridges.

Halogen lamps are an improved version of incandescent lamps, with better average life and a smaller bulb.

12.2 Types of lamps

Halogen lamps come in a rather wide variety of different types and shapes, still it is useful to learn to identify a few of them.

According to voltage, halogen lamps can be either normal tension or low tension (this last group requires a transformer). Low voltage luminaires are the only type allowed in certain areas of bathrooms, because of the risk of electric shock.

Concerning the shape, there are two main groups: with reflector and without.

12.2.1 Lamps without reflector

There are two main groups of these (Figure 69):

1. Single ended, among these we note lamps with an E14 or E27 cap, whose function is to replace traditional incandescents,
2. Double ended lamps: usually installed in floodlights.

Halogens were the first compact high quality lamps. In particular the low-tension halogen first used in headlamps found a new life in table lamps. From the beginning of the seventies, with the world famous Tizio table lamp by R. Sapper for Artemide, it was the symbol of a new futuristic design coupled with technical innovations. The lamp does away with wires, because it is powered through the rods, connected together by press-button joints and carefully counterbalanced.

Figure 71: Cone diagram for Philips Masterline ES halogen lamp.

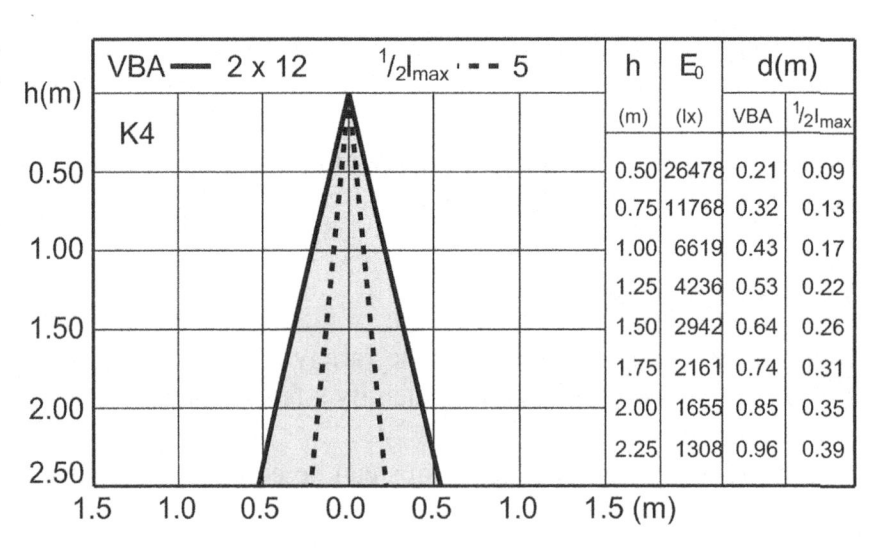

h (m)	E_0 (lx)	d(m) VBA	d(m) $^1/_2 I_{max}$
0.50	26478	0.21	0.09
0.75	11768	0.32	0.13
1.00	6619	0.43	0.17
1.25	4236	0.53	0.22
1.50	2942	0.64	0.26
1.75	2161	0.74	0.31
2.00	1655	0.85	0.35
2.25	1308	0.96	0.39

12.2.2 Lamps with reflector

Lamps with reflectors (Figure 70) are generally used in cheaper lower-quality spotlights.

The reflector gives shape to the luminous flux and sends it towards the axis of the lamp with a specific beam spread. To use them the designer must not only select the power, but also the beam spread that satisfies the requirements.

Just like spotlights, these lamps have their cone diagram (Figure 71).

12.2.2.1 *Halogens with dichroic reflector*

A major disadvantage of the halogen lamp is the significant amount of heat it produces, and, in the special case of lamps with reflector, concentrates on the task.

In order to reduce this problem the dichroic reflector (such as the one in the top left lamp in Figure 70) was introduced: it is a glass reflector treated so that it is transparent to infrared radiation and specular to visible light. Thanks to that, any infrared emitted by the lamp passes through the glass, so that it can be dispersed all around it and not concentrated onto the task.

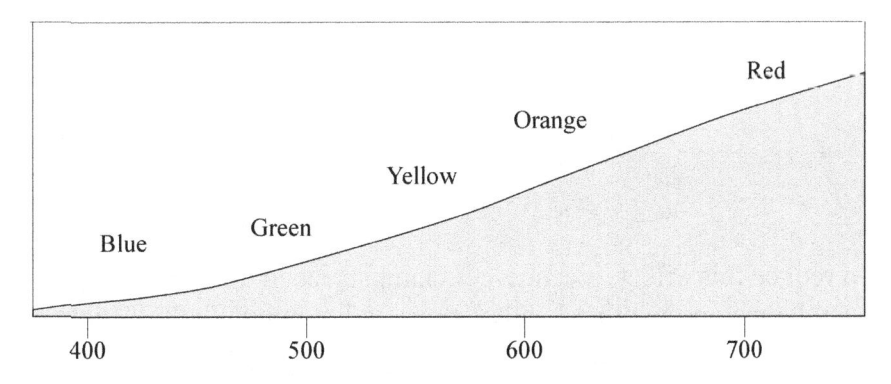

Figure 72: Spectrum of a halogen lamp.

12.3 Description

The halogen lamp is an improvement on the incandescent, it works in a very similar way and it emits a very similar spectrum of light.

Let's look at Figure 72: the emissions in the red are much higher than those in the blue, and this trend will continue in the infrared, beyond the visible range. This shows how most of the energy absorbed by the lamp is actually turned into infrared radiation, or heat, and it explains the very low efficacy of this type of source. It also explains why halogen lamps can only emit warm light.

These lamps have high luminance, so they must either be screened or aimed away from the lines of sight of the users.

12.3.1 Inner workings and halogen cycle

A tungsten wire is made incandescent by the electric current (Figure 73) coursing through it. In order to reduce heat dispersions and to reach even higher temperatures, the wire is wound up in a spiral. The wire is contained in a quartz bulb resistant to high temperatures.

The main disadvantage of having a white-hot tungsten wire inside a bulb is that atoms of the wire, excited by the heat, will leave the wire and will coat the inner surface of the bulb. This blackens it, and significantly reduces the amount of light exiting the lamp.

Figure 73: Halogen lamp, with tungsten wire and quartz bulb.

To reduce this effect, incandescent lamps needed a large bulb. Halogen lamps on the other hand manage with a minute bulb thanks to the halogen cycle. As we know, a mixture of halogen gases fills the bulb; these form a chemical bond with the atoms of tungsten that have left the wire, and stop the deposition of metal on the surface of the quartz. When the tungsten-halogen molecule is brought back, by convection, towards the wire, the high temperature breaks the bond, and some of the tungsten returns onto the wire.

This cycle has the added effect of slightly prolonging the average rated life of the lamp.

12.4 Evaluating grid

Let's apply our evaluating grid to this source (Table 17).

Indicator	Performance
Bulb and cap	many, small to medium
Colour temperature	from 2800 K to 3200 K
Colour rendering	CRI = 100
Luminous efficacy	Low
Average life	from 2000 to 5000 hours
Electrical devices	both with and without

Table 17: Evaluation of halogen lamps.

12.4.1 Bulb and cap

Halogens come in a wide variety of sizes and shapes, from the glass bulb of the replacement for the now retired incandescent, to the minute quartz bulb of the low tension versions, to the PAR lamps, i.e. those with integrated reflectors. Figure 69 and Figure 70 try and document the range of shapes and sizes. Still, compared to the fluorescent in particular, the general size of this kind of lamp is small to medium.

12.4.2 Colour temperature

Because of the incandescence process, the correlated colour temperature for this lamp will be warm, up to about 3000 K.

12.4.3 Colour rendering

The colour rendering index for halogen lamps is the highest available. Please keep in mind the considerations made in "2.5 Colour rendering" on page 22.

12.4.4 Luminous efficacy

Luminous efficacy is so bad that the change in regulations that caused the retirement of the incandescent decimated the range of halogen lamps as well. The manufacturers had to upgrade the technology slightly in order to increase the efficacy over the limit imposed by the new regulations. Even with the upgrade, luminous efficacy is the lowest.

12.4.5 Average rated life

Thanks to the halogen cycle, the average rated life varies from about 2000 to 5000 hours. Unfortunately it is the shortest among the available types of lamp.

12.4.6 Electrical devices

Some halogen lamps require a transformer (low tension), others connect directly to the mains. Choosing among the two is mainly a matter of safety: low tension lamps eliminate the risk of electrocution.

12.5 Considerations

12.5.1 Strengths

The most important advantage of halogens is surely their very high colour rendering index.

The warm colour temperature is useful to create a cosy atmosphere in the interior.

Historically, the smaller size has helped significantly with their diffusion, but today, with the advent of LED technology and the improvement in metal halides, there are better solutions where miniaturisation is necessary.

Halogen lamps

Halogen are the cheapest lamps to purchase, thanks to a relatively simple and very mature manufacturing technology. We shall see later on that this does not mean they are the cheapest to run.

12.5.2 Weaknesses

Obviously the most damaging limit of this technology is its inherent low efficacy. When they appeared, their quality of light o was so much higher than the rest that it made halogen the right choice if one wanted good light, and fluorescent the choice if one wanted cheap light. Nowadays the quality of light of the other type of lamps has improved so much that it is quite hard to recommend a halogen lamp.

The short rated life does not help either; it means that selecting a halogen will have the client facing higher maintenance costs.

12.6 Examples of fixtures

Historically the most important fixtures in this group were table lamps or spotlights intended to illuminate merchandise for which high colour rendering is ideal. The cheapest spots use lamps with reflector, the higher quality ones use only naked lamps, and provide their own reflector.

Figure 74: Example of light fitting using halogen lamps.
On the left a surface mounted projector, in the centre a recessed grid-light with four independent lamps, on the right a table lamp for medical use.

A relatively frequent field of application for these lamps used to be the lighting of art. When lighting the most delicate items, it is important to make sure that the fixture contains an UV filter, even though the ultraviolet emission of the halogen lamp is quite minor.

12.7 Self evaluation

Answer the following questions to test your understanding of the material:

1. Describe the characteristics of incandescent lamps and why they are being retired.
2. List the characteristics of the halogen lamp that contribute to the evaluation table.
3. Describe the strengths and weaknesses of these lamps.
4. Explain what is the halogen cycle.
5. Give examples of light fittings that employ halogens and of their fields of application.

Figure 75: Most common shapes of traditional fluorescent lamps, drawn to scale.
From left to right: a linear fluorescent, a ring shaped lamp and a black light lamp.

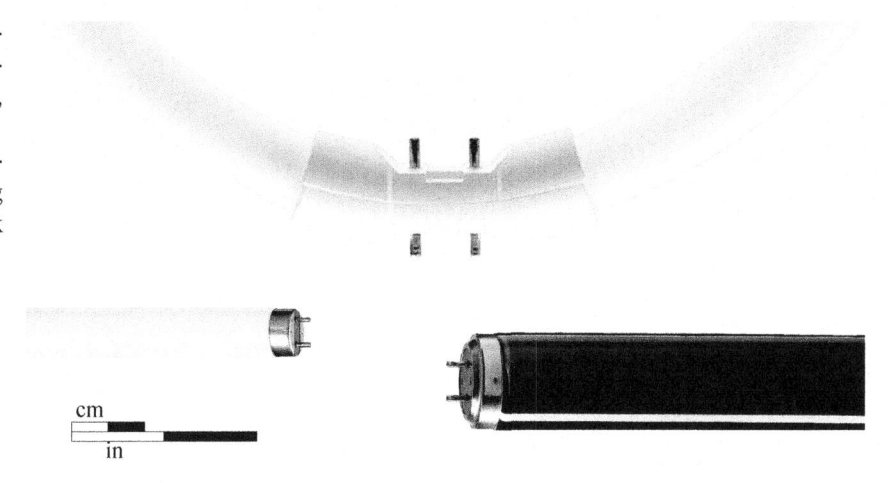

Figure 76: Examples of compact fluorescent integrated lamps, drawn to scale.
On the left: lamps without reflector, on the right one with it.

Figure 77: Examples of compact fluorescent non-integrated lamps, drawn to scale.

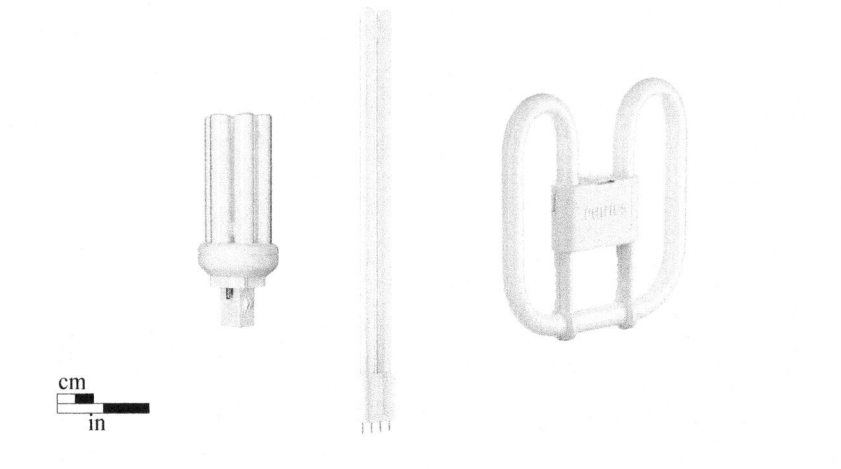

Fluorescent lamp

13

Fluorescent lamps come in a very rich variety of shapes: from the classic tube, to lamps that could pass for frosted incandescent. They produce light with the widest range of colour rendering and temperature.

13.1 Lamp types

There are two main types of lamps: traditional fluorescents, and compact fluorescents.

Traditional fluorescents (Figure 75) come in three basic shapes: Tube, U shaped (not available in Europe) and ring shaped.

Compact fluorescents have many forms, and their common characteristic is the smaller (for fluorescents) size. They can be integrated (Figure 76), which means they work connected directly to the mains or non-integrated (Figure 77), which means that they need external electrical devices (see 13.3.6) to work. The most widely known integrated compact fluorescent lamps are those commonly called "energy-saving": they were the first attempt to find a more efficient solution that could replace incandescent lamps.

These lamps are rather large, this means that their luminance is low, and, because of that, even when they are used without diffusing screens they will hardly damage visual comfort. This allows the fixture to have a better R_{LO} (4.4 on page 43).

13.2 Description

Figure 78 shows the spectrum of emission of a fluorescent lamp: it is not continuous like the halogen's, but it shows multiple discrete bumps where the different phosphors emit light. It is easy to deduce that the quality of light will depend on how well the phosphors work.

The quality and number of phosphor compounds used in the coating has significantly increased in the last forty years. Because of that, the historical limitations of this kind of technology (bad colour rendering, a very limited range of colour temperatures, and flicker) have more or less disappeared.

Figure 78: Spectrum of a fluorescent lamp.

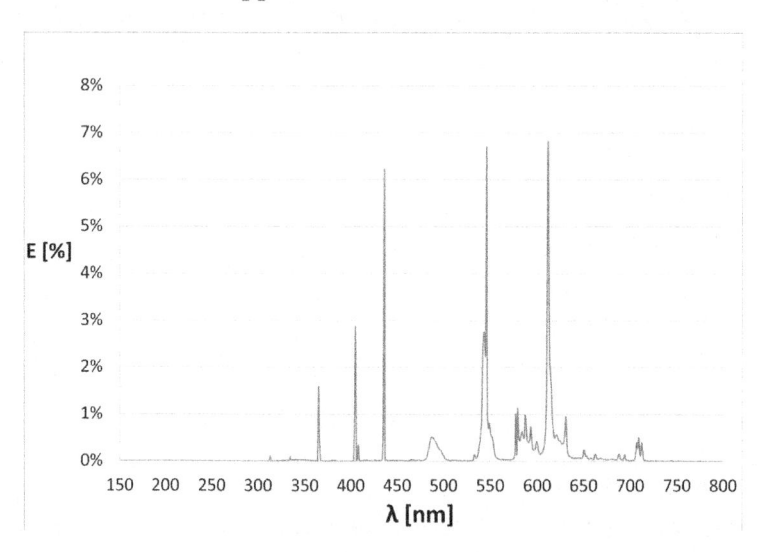

In the next few sections I will paint a more contemporary picture of the situation. Just looking at Table 18 the reader can see how many combinations of colour temperature and colour rendering index are available today.

Table 18: Versions of fluorescent lamp available, ordered by colour temperature and colour rendering index.

	\multicolumn{7}{c}{Correlated colour temperature}
CRI	2700 K	3000 K	3500 K	4000 K	5400 K	6500 K	8000 K
90-100		●		●	●	●	
80-89	●	●	●	●		●	●
70-79			●	●		●	
60-69				●			

13.2.1 Inner workings

The discharge tube of the lamp contains mercury gas at very low pressure. The electric arc that goes through it excites the gas, which produces UV radiation. The inner surface of the tube has a coating of phosphor dust: once the ultraviolet radiations hit it, they are absorbed and the coating emits light.

13.3 Evaluating grid

Let's apply the evaluation grid to this technology.

13.3.1 Bulb and cap

Traditional fluorescents are large tubes, either straight, or curled up in a "U" or "O" shape. This is a limit on their application, because they force the shape of the fittings into certain silhouettes, and don't allow for small luminaires.

Compact fluorescents come in a much larger variety of shapes. Their size is smaller, the smallest is as big as a medium to large halogen.

Caps for integrated compact fluorescents are those found in incandescents and halogens, as they work directly connected to the mains. All the rest of the fluorescents require electrical devices, and the caps are different. Sometimes the same lamp can have a different cap whether it is used with a traditional or electronic ballast.

13.3.2 Colour temperature

Because of the flexibility of the phosphor coating, fluorescents emit light in a very wide range of colour temperatures: from 2200K to 8000K (Table 18).

13.3.3 Colour rendering

The performance here goes from good to excellent: the worst CRI available on the market today is sixty (for a very limited number of legacy lamps), and the best is ninety (Table 18).

13.3.4 Luminous efficacy

Fluorescents produce light using two very efficient phenomena, so the end result must be very good. We shall see that the subtypes have similar results: slightly better for traditionals, and slightly worse for compacts (in particular the integrated ones).

13.3.5 Average rated life

Average life is rather long: up to about 18000 hours.

13.3.6 Electrical devices

Because of the electric arc, these lamps need both a starter and ballast. Normally the electronic ballast is chosen, which contains both.

The compact integrated fluorescents have ballast and starter integrated into the cap.

Table 19: Evaluation table for fluorescent lamps.

Indicator	Performance
Bulb and cap	medium to large
Colour temperature	from 2200 K to 8000 K
Colour rendering	from 60 to over 90
Luminous efficacy	high
Average life	around 18000 h
Electrical devices	always (sometimes integrated)

As usual, the electronic ballast is preferable:

1. It brings the lamp to full functioning power more quickly,

2. It is able to constantly monitor the behaviour of the lamp and keep it on its optimal conditions,
3. It stops trying to turn on the lamp when it has failed (no risk of burning the ballast),
4. It removes the problem of flickering that has plagued these lamps since their inception.

13.4 Traditional Fluorescent lamps

Figure 75 shows the shapes for traditional fluorescent lamps: the ring is frequently used for kitchen pendants, while the tube is surely the most familiar. The tube's diameters vary from 16 mm to 26 mm, and their length from around 220 mm to over a metre and a half.

13.4.1 Considerations

The excellent luminous efficacy, the high colour rendering index, and the long average life make the fluorescent lamp a very good choice.

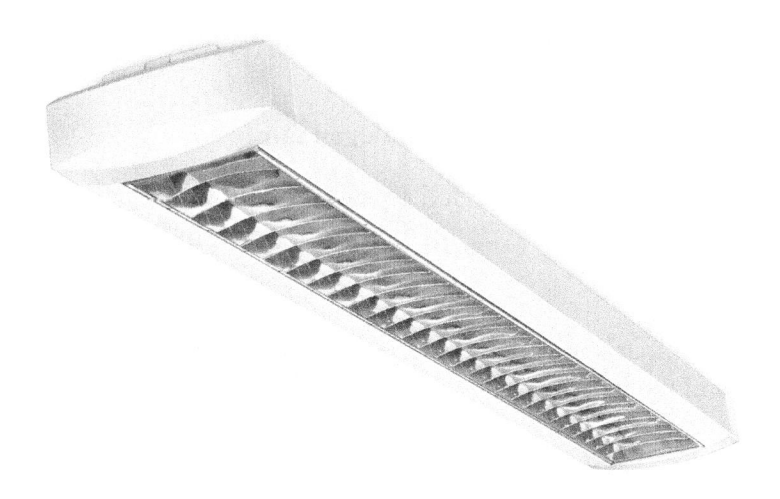

Figure 79: Example of light fitting with linear fluorescent lamp.

The main disadvantage is the large size: this means that these lamps cannot be used in small fittings, and that the exterior shape of the luminaire needs to more or less follow the shape of the lamp.

13.4.2 Example of fittings

Light fittings with fluorescents can look less interesting, and more technical.

Their shape (see Figure 79) is some sort of elongated box that contains the lamps, usually ceiling mounted, suspended, or in a line system.

13.5 Compact integrated fluorescent lamps

Compact fluorescents are an attempt to provide the typical efficient and high-quality light of the fluorescent without the large size of the traditional lamps.

This type of lamp (Figure 76) has an electronic ballast miniaturised into the base. Thanks to that, it can replace some traditional incandescents (or halogens) without modifying the fitting in any way. The vast majority of these lamps, with or without reflector, has an Edison cap, either E14 or E27, because these are the types of caps that fit into the traditional lamp holders for incandescents.

13.5.1 Cost comparison to the halogen lamp

Since these lamps were created to save energy, it makes sense to compare them to other types of light sources from an economical perspective.

The first comparison I would present is between the compact fluorescent and the halogen.

The hypotheses are:

1. Electricity costs: 0.3 Euro per kilowatt-hour.
2. Replacing a lamp costs five Euros.
3. Lamps are replaced at two thirds of their average rated life.
4. A work-day for each lamp is eight hours long.

Fluorescent lamp

I am going to compare lamps that share colour temperature, colour rendering index, emit the same luminous flux and have the same overall shape and size.

It is immediately apparent (Table 20) that the advantages of the compact fluorescent lamp are the power absorbed, and the average life; while those of the halogen are the purchasing cost and the colour rendering index.

Image	Brand	Model	W	Avg. Life h	Cost €	Flux lm	CCT K	CRI
	Philips	Ecoclassic	53	2000	2	850	2700	10
	Philips	Softone	16	12000	33.54	860	2700	8

Table 20: Cost comparison between halogen and compact fluorescent.

The question becomes whether fewer replacements and low consumption will balance out the much higher purchasing cost.

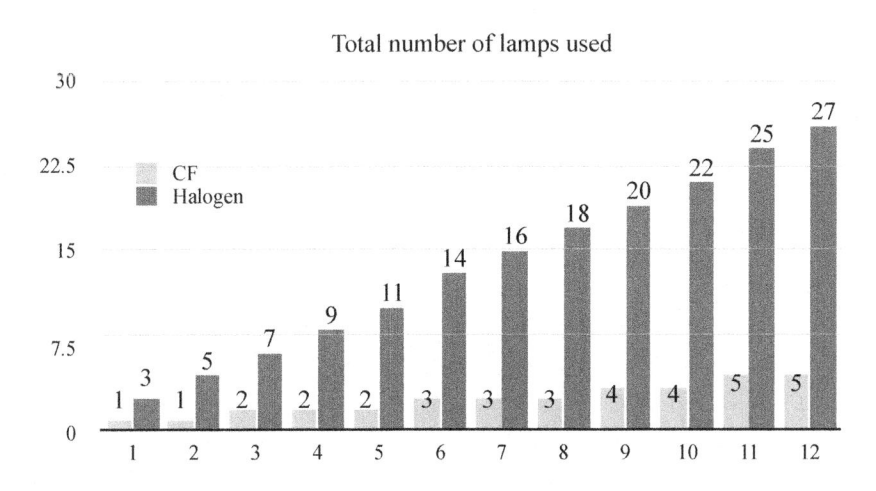

Figure 80: Total number of lamps used per year, in light grey compact fluorescents, in dark grey halogens..

Figure 80 shows the number of lamps necessary over time (years), and it helps explaining why longer life and lower energy consumption become more important than purchasing cost: at the end of the third year the installation with a compact fluorescent is still using its second lamp while the installation with a halogen is using its seventh, and all this time the compact fluorescent has absorbed about a third of the power absorbed by the halogen.

Figure 81: Savings obtained by replacing a halogen with a compact fluorescent, each year.

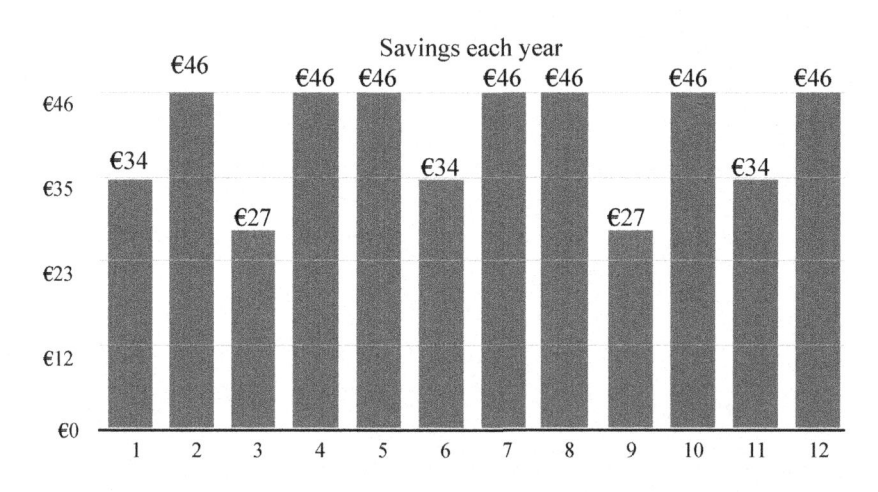

In a short time, the saving becomes considerable: more than a hundred Euros in the third year per lamp (Figure 82).

Figure 82: Total savings selecting the compact fluorescent instead of the halogen, year after year.

Figure 81 shows that the saving is periodic, with a minimum at 27 € when the compact fluorescent needs to be replaced.

13.6 Compact non-integrated fluorescent lamps

Non-integrated compact fluorescent lamps (Figure 77) work exactly like traditional ones: the only difference is that the lamp contains only the cap, the discharge tubes, and, in some cases, the bulb. The ballast and starter (or the electronic ballast) are in the fitting.

The advantage of non integrated lamps is that, since there is no need for miniaturisation, ballast and starters will be more efficient and live longer.

13.6.1 Considerations

Compact fluorescent conjugate smaller size with high quality of light and high efficacy; unfortunately the size of these lamps is still considerable, and it is what limits their flexibility the most.

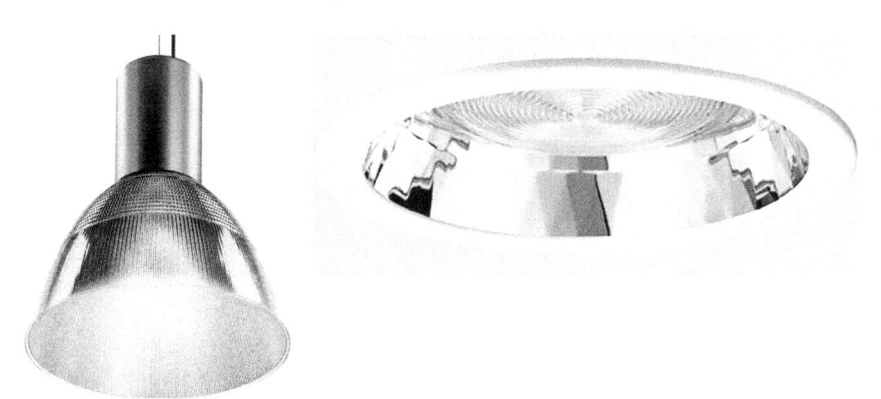

Figure 83: Examples of fittings with compact fluorescent lamps (not to scale).

13.6.2 Example of light fittings

Figure 83 shows two types of fixtures that use compact fluorescent lamps: they are smaller and, in some cases, they can be open and show the lamp: this increases the light output ratio without causing glare because of the low luminance of the source.

13.7 Self evaluation

Answer the following questions to test your understanding of the material:

1. Describe the types of fluorescent lamps, detailing their differences and their similarities.
2. List the characteristics of the fluorescent lamp that contribute to the evaluation table.
3. Describe the strengths and weaknesses of these lamps.
4. Explain how do they work and the consequences on their performance.
5. How expensive is it to purchase and use a compact fluorescent, compared to an halogen?
6. Give examples of light fittings that employ each type of fluorescent and of their fields of application.

Figure 84: Examples of metal halide lamps, both with and without reflector, drawn to scale. Notice that the smallest lamp is close in size to a medium sized halogen.

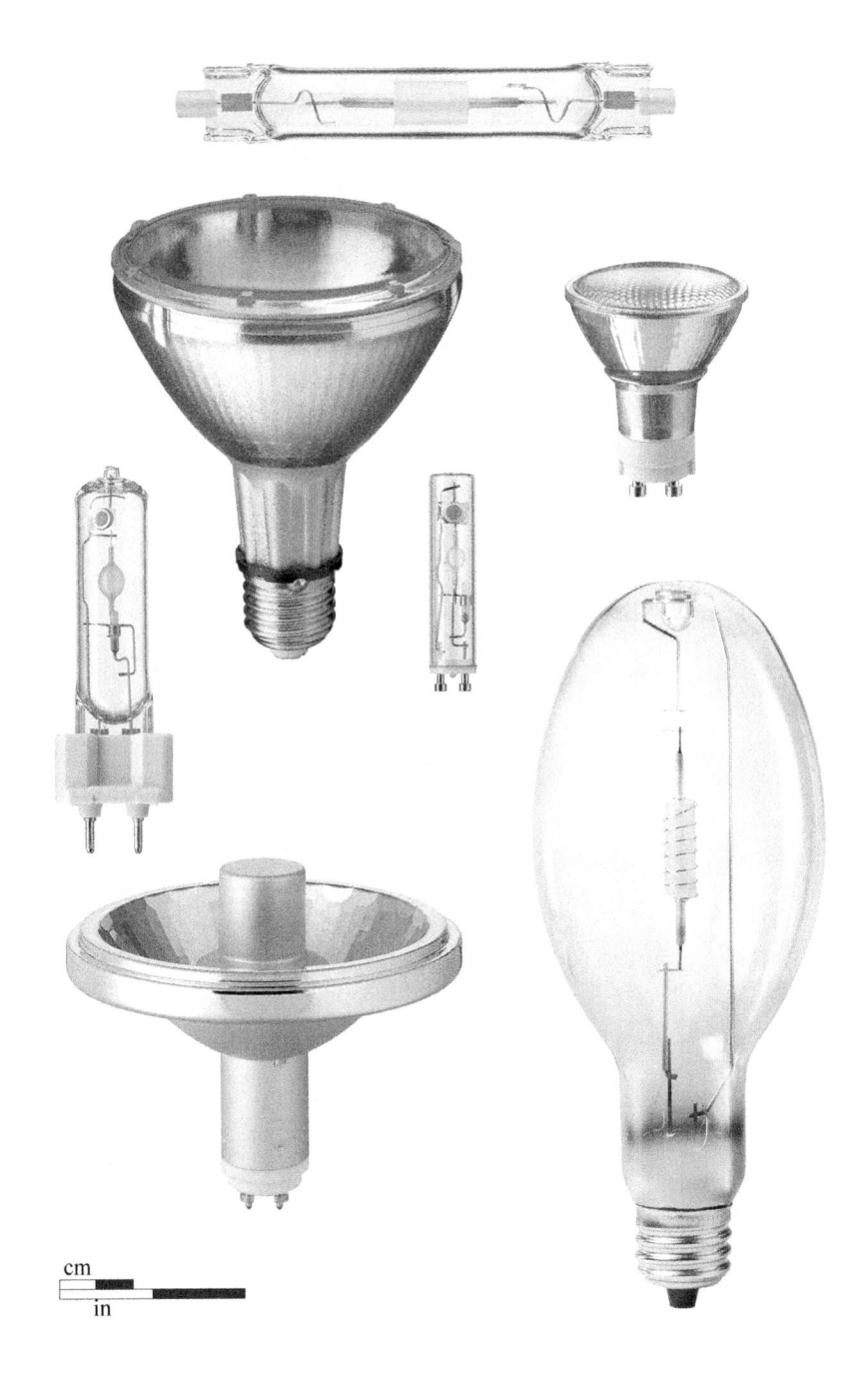

Metal halide lamps 14

In general, discharge lamps are large and very powerful, they can absorb up to 2000 W; and are used for outdoor areas (e.g. sport arenas).

Compact discharge lamps are their smaller cousins and, among them, only metal halide lamps are used indoors. They produce a high quality light, and they are very efficient.

14.1 Types of lamps

Just like halogens, metal halides come in three main shapes:

1. Without reflector and single ended
2. Without reflector and double ended
3. With reflector.

Figure 84 shows a few metal halides with and without reflector: it is interesting to compare it to Figure 69 on page 134 and Figure 70 on page 134: the smaller metal halides are almost as big as the larger halogens.

14.2 Inner workings

This type of lamps produces light when an electric arc passes through a mixture of gases, inside a quartz or ceramic arc tube. This must be able to withstand very high temperatures (over 600 °C), and it is protected from interaction with the external environment by the lamp's bulb. In the arc tube there is a specially prepared mixture of metal halides (compounds of metals and halogens), rare earths, and mercury. When the electric arc passes through the tube, the gases turn into plasma, and produce light (Figure 85).

Figure 85: Spectrum of a metal halide lamp (Philips CDM-T Evo).

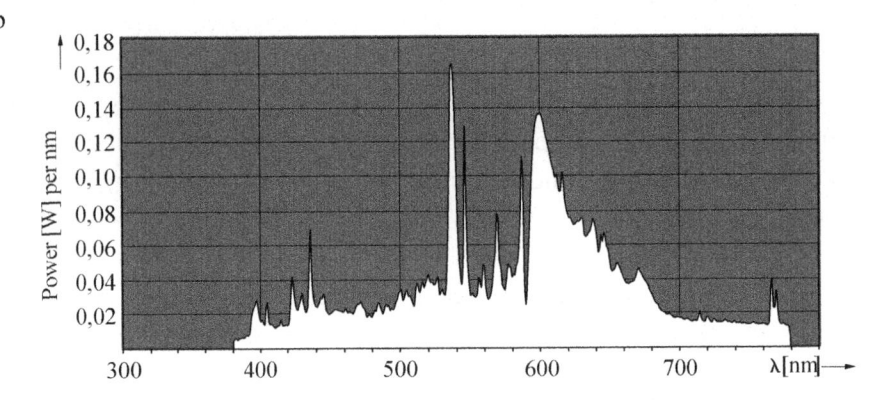

Mercury produces visible light (blue part of the spectrum) but also UV radiation, which is usually screened by the outer bulb of the lamp. Because UV is so dangerous, it is important to verify that the lamp bears the "UV STOP" symbol, especially when lighting delicate items. In this last case it is safer to employ a fitting equipped with a UV filter as well.

14.3 Totally enclosed luminaires only

The interior of the arc tube is a formidable environment: high temperatures, major chemical aggression from the compounds and extreme pressure. Ceramic materials are employed for their higher resistance to temperature, pressure and chemical attack, but there is always the remote possibility of a catastrophic failure.

Because this lamp reaches such a high temperature (600 °C in the arc tube, up to 500 °C on the bulb and about 250 °C on the cap), it is mandatory to select a totally enclosed fitting (even during testing) and one capable of containing hot lamp parts. This ensures safety in the ultra rare but still dangerous event of a non-passive rupture.

14.4 Start up and warm restart time

Even though the electric arc produces inordinate amounts of heat, it takes a little time for the plasma in the tube to form and produce light. This is the reason why, when cold, the metal halide lamp will take about a couple of minutes to reach full power, and full colour rendering. As the different compounds in the arc tube light up, it will show multiple flashes of colours.

When the lamp is turned off then immediately on again, the temperature and pressure inside the tube are so high that the normal voltage from the ballast is not enough to generate the arc. The lamp will be able to turn on again only after it has cooled itself slightly, in about five to ten minutes.

This is the biggest limitation of this otherwise very performing technology.

14.5 Cyclic behaviour at end of life

At the end of the life of the lamp, the voltage required to sustain the electric arc increases beyond the capability of the ballast. The lamp can be turned on, the arc will start warming up the gases into the tube, but soon the voltage will be insufficient, the arc will cease, the chamber will cool down, the voltage will again be sufficient for the arc and the process will begin anew.

This loop can cause both the catastrophic failure of the lamp and the burning out of the ballast and starter.

The electronic ballast recognises this behaviour and stops the lamp from turning on when it is exhausted.

14.6 Cost comparison to the halogen lamp

I am going to introduce again the cost comparison between an A55 halogen lamp and a metal halide one. Table 21 contains the lamps' data.

Table 21: Cost comparison between halogen and compact fluorescent.

Image	Brand	Model	W	Avg. Life h	Cost €	Flux lm	CCT K	CRI
	Philips	Ecoclassic	53	2000	2	850	2700	10
	Philips	CDM-T	20	15000	13	1800	3000	8

It is evident how even the least powerful metal halide lamp emits more than twice the luminous flux of the halogen. So, in order to compare flux to flux we need to use two of the latter for each of the former.

The same hypotheses of electricity cost (0.30 Euro per kWh), maintenance cost (five Euro) and maintenance interval (two thirds of the average rated life) apply.

The metal halide lamp's cost is three times that of the halogen (considering two lamps) and the average life is seven and a half times longer!

Figure 86 shows the effect of the average life: at the end of the fourth year, when the first couple of metal halide lamps replacement happens, the halogens have been replaced eighteen times.

Figure 87 describes the significant total savings, and Figure 88 shows how cost effective is the metal halide compared to the halogens. The difference is staggering: it goes from a minimum of 84 € per year per lamp to a maximum of 116 €, with the running total reaching almost 800 € in eight years.

Metal halide lamps

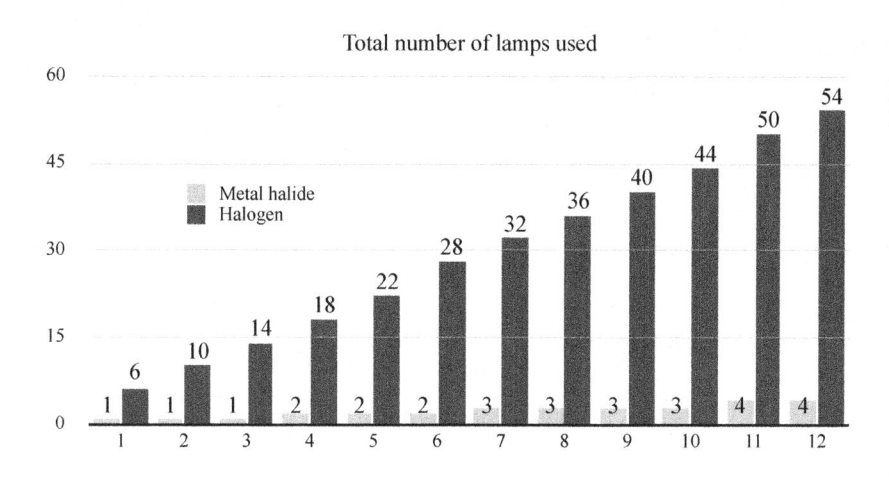

Figure 86: Total number of lamps used per year. In blue the metal halides, in green the halogens.

Figure 87: Total savings replacing the halogen with the metal halide, year after year.

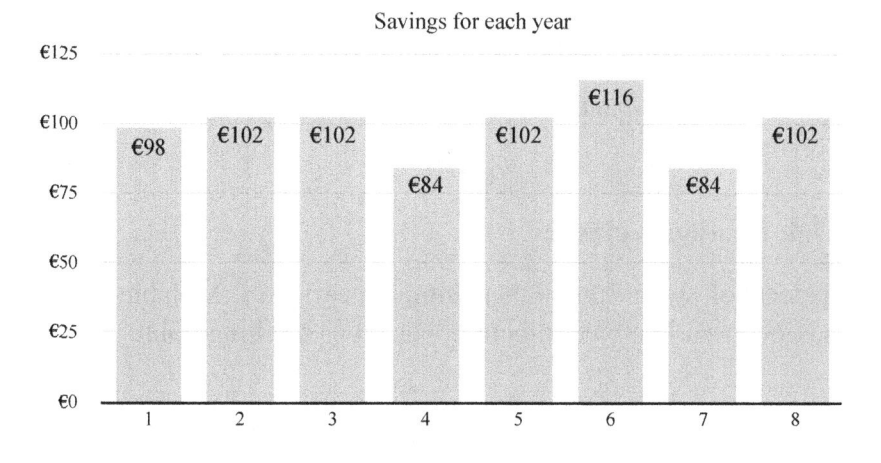

Figure 88: Savings selecting the metal halide instead of the halogen, each year.

14.7 Evaluating grid

It is shown in Table 22.

Table 22: Evaluation table for metal halide lamps.

Indicator	Performance
Bulb and cap	small to medium
Colour temperature	from 2500 K to 6000 K
Colour rendering	over 80
Luminous efficacy	very high
Average life	from 9000 to 20000 h
Electrical devices	always

14.7.1 Bulb and cap

The size of the bulbs go from small to medium. It means the lamp can be employed in all but the smallest fittings.

14.7.2 Colour temperature

These lamps, while not having the flexibility of fluorescents, can still produce warm, neutral, and cool light. The actual colour temperature range goes from 2500 K (for special lamps) to 6000 K.

14.7.3 Colour rendering index

Metal halide lamps have excellent colour rendering, ranging from 80 to about 94.

14.7.4 Luminous efficacy

This technology has the highest luminous efficacy available now. It is so good that lamp manufacturers use it as a selling point.

14.7.5 Average rated life

The average rated life is the second best among all the lamp technologies that we will look into. Again it is such a good value that manufacturers publish it in the catalogues: ranging from a minimum of 9000 h to a maximum over 20000 h.

14.7.6 Electrical devices

Being a discharge lamp obviously means that it will need either a ballast and a starter to work, or just an electronic ballast.

Using an electronic ballast is a particularly good idea with metal halides, because:

1. It can control both the current and the voltage applied to the lamp, so that it is always working at it as best efficiency,
2. It gives quicker cold and warm start-ups,
3. It makes the lamp live longer,
4. It removes flickering,
5. It recognises and interrupts the end of life cyclic behaviour before damaging the system.

14.8 Considerations

The final evaluations of this kind of technology is absolutely excellent, apart from the major disadvantage of the long start up time.

14.8.1 Strengths

Among the advantages there are the average rated life, the excellent colour rendering, the wide range of colour temperatures, the small size, and the high luminous efficacy.

14.8.2 Weaknesses

The most important disadvantage is surely the start-up time, because it strongly limits the fields of application of this lamp. It is excellent for generally lighting, or highlighting in situations where high luminous flux is required, but it must be an installation where the lamps are turned on and off very rarely.

Even if they tend to be pricey (over 100 € for special sources, about ten for normal ones), they turn out to be cheaper them halogens.

It is important to always keep in mind the need to select completely enclosed and resistant fittings, and to check the existence of ultraviolet filtering on the lamp.

14.9 Example of light fittings

Figure 89: Application of a luminaire powered by a metal halide lamp (Philips Mini 300 Cube) to general lighting with a high ceiling and wide areas,

Metal halides are used in powerful spotlights and downlights, or projectors. These are usually employed to illuminate large areas with high ceilings (downlights and projectors) or to highlight items (spotlights) from very far away. Figure 89 shows an example.

Metal halide lamps 165

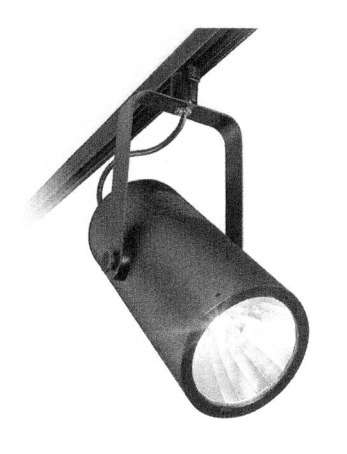

Figure 90: Luminaires powered by a metal halide lamp (Philips Store-Fit projector and Store-Fit downlight)

Figure 90 shows an example of track mounted spotlight (left) and an aim-able recessed downlight.

Figure 91: Application of a luminaire powered by a metal halide lamp (Philips Mini 300 Stealth)

Figure 91 shows a row of indirect luminaires that illuminate a distant glass and steel ceiling.

14.10 Self evaluation

Answer the following questions to test your understanding of the material:

1. Describe the characteristics of the metal halide lamp that contribute to the evaluation table.
2. Describe the strengths and weaknesses of these lamps.
3. Explain how do they work and the consequences on their performance.
4. How expensive is it to purchase and use, compared to an halogen?
5. Describe the typical application and give examples.

Figure 92: Example of LED lamps without reflector.

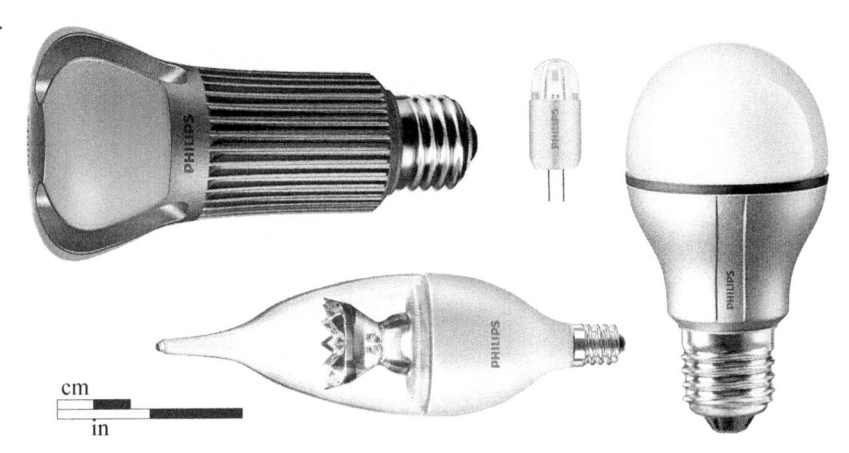

Figure 93: Example of LED lamps with reflectors.

Figure 94: Examples of LED modules.

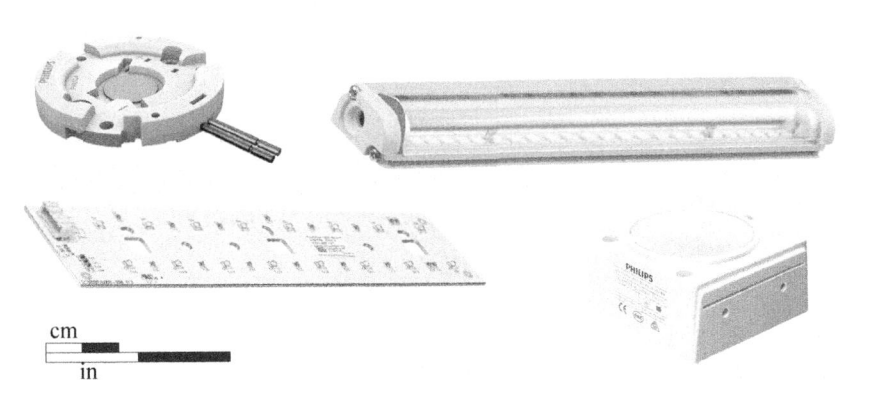

LEDs

15

LED light sources have taken over the market: every manufacturer has to have them, and there is a rush to try and use them everywhere. In this chapter I will try to separate the marketing from the technical, and give a more sober presentation of the LED lamps.

There are three major types of LEDs available on the market:

1. Signal LEDs: they are monochromatic, and they have been available for many decades: an example is the red dot in television sets. They do not interest us.
2. LED lamps (Figure 92 and Figure 93): they aim to replace traditional light sources. It is the more recent application of this technology.
3. LED modules (Figure 94): both with coloured and white light.

15.1 LED lamps

LED lamps were invented in order to offer a new replacement for traditional sources, when the technology reached the needed efficacy and light quality. Figure 92 and Figure 93 shows a few examples of these relatively new lamps: there is a replacement for a dichroic halogen, for a E14 and E27 incandescent and, weirdly enough, a replacement for a fluorescent tube (not in the picture).

Replacing a fluorescent with a LED is ambitious: fluorescent lamps have very high light quality, low luminance and emit diffused light from an extended source; LEDs are the exact opposite: light quality isn't outstanding (on average), their luminance is very high, the emission is all but diffused and the sources are tiny.

Replacing an incandescent lamp is much more feasible.

15.1.1 Comparison to halogens

I am presenting the usual comparison: a halogen lamp (Philips Eco-Classic) and a LED lamp that emits more or less the same luminous flux.

Table 23: Cost comparison between halogen and LED.

Image	Brand	Model	W	Avg. Life h	Cost €	Flux lm	CCT K	CRI
	Philips	Ecoclassic	53	2000	2	850	2700	10
	Philips	LEDbulb	10	25000	19.55	860	2700	8

I maintain the same hypotheses: an eight-hour day, electricity cost at €0.30 per kilowatt-hour, maintenance cost at five Euros per lamp, and replacement at two thirds of the average rated life.

The luminous flux emitted by the two lamps is about the same (there's a 6% difference).

The LED lamp absorbs a fourth of the power, has twelve and a half times the average life, and costs ten times as much as the halogen.

The effects of the very long life of the LED lamp is shown in Figure 95. There we can see that when the LED lamp is replaced (before the end of the sixth year of use), the halogen lamp has been already replaced fourteen times.

The cumulative effect of the longer life, lower energy consumption, and fewer maintenance cycles is shown in Figure 96, where the total saving of the LED solution is displayed.

LEDs

Number of lamps

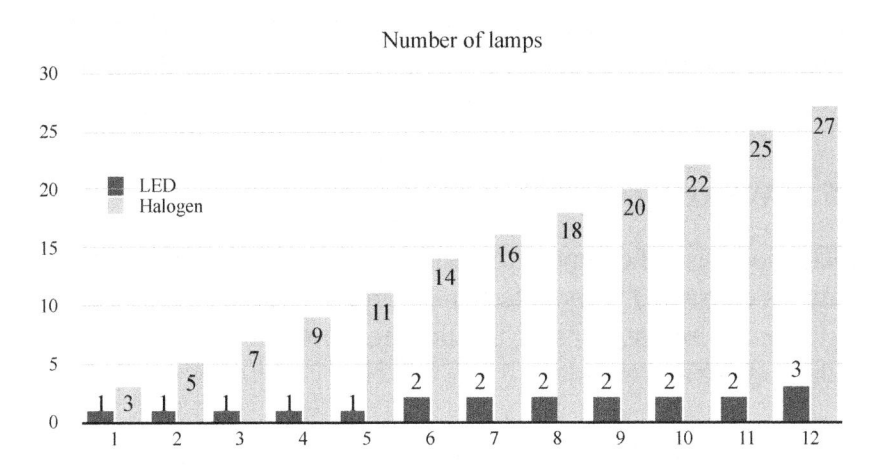

Figure 95: Total number of lamps used per year. In blue LEDs, in green halogens.

Total savings from using the LED lamp instead of the halogen

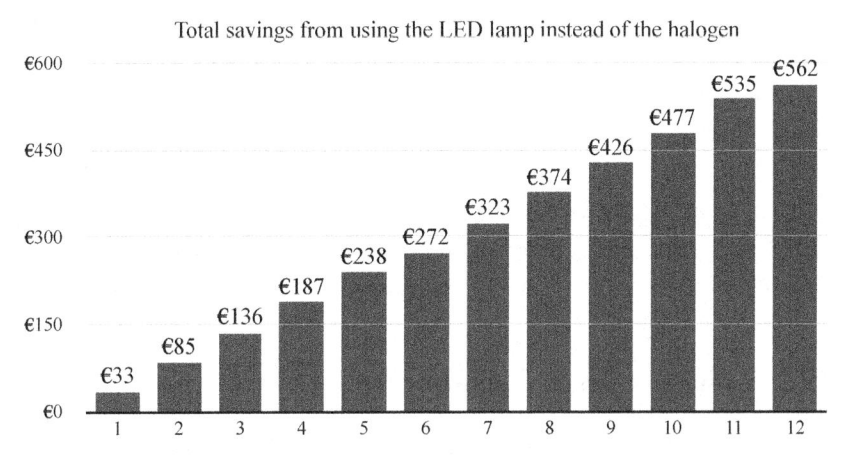

Figure 96: Total savings replacing the halogen with the LED, year after year.

Savings per year

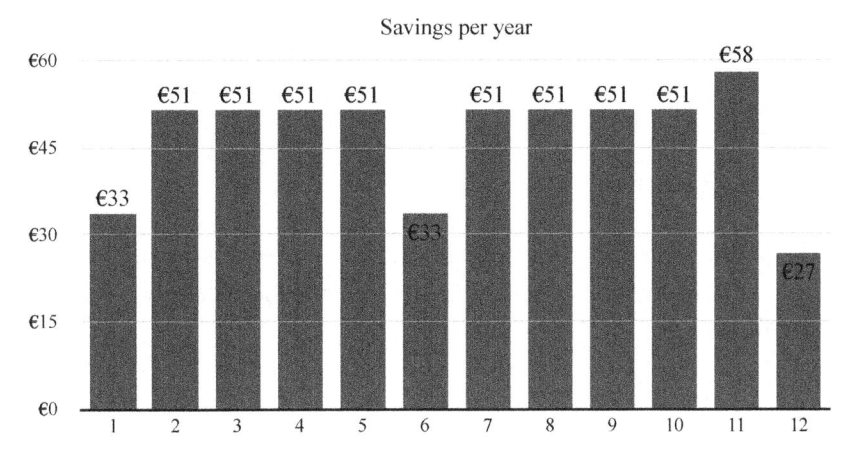

Figure 97: Savings each year, selecting the LED instead of the halogen, year after year.

Not surprisingly, the LED lamp is much cheaper than the halogen, with savings per year (Figure 97) cycling from 27 € to 58 €, and total savings (Figure 96) reaching 477 € after ten years.

15.2 White light LED modules

LED modules are composed by the LEDs (warm or cool light) each with its own PCB[1], mounted on a support. These supports can be tailor-made into any shape, but the most common ones are rectangular, square and circular.

Almost any number of LEDs, with any orientation, can be placed on a support (Figure 94 on page 168): normally a square support will have a grid of LEDs , a narrow rectangular (strip) support one or more rows of LEDs, and a circular support a radial disposition of LEDs. The supports can be rigid, flexible or a mixture of the two.

A diffuser or a lens system can be mounted over the support to reduce luminance and shape the light, and a metal profile should be connected under the support to provide heat dissipation.

15.3 Coloured light LED modules

Coloured light LED modules have exactly the same structure of white light ones, with the obvious difference of providing coloured light either by using RGB LEDs that contain multiple emitters, or by mixing different monochromatic LEDs.

15.4 Description

LED stands for light emitting diode: an electronic semiconductor device that allows the passing of current in one direction only, and emits light.

---

1. Printed Circuit Board.

15.4.1 Inner workings

A typical LED is composed of:

1. A transparent silicone resin dome, it protects the emitter and shapes the light,
2. A semiconductor emitter that produces light,
3. A ceramic support with a metal structure within for heat conduction.

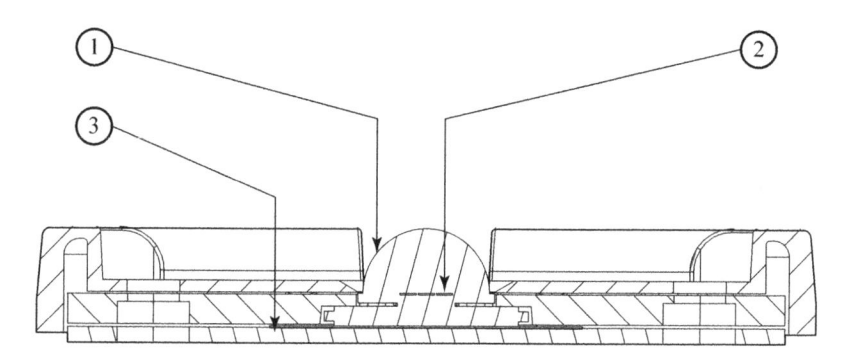

Figure 98: Scheme of a LED.
1: Resin dome
2: Emitter
3: Ceramic support with metal core.

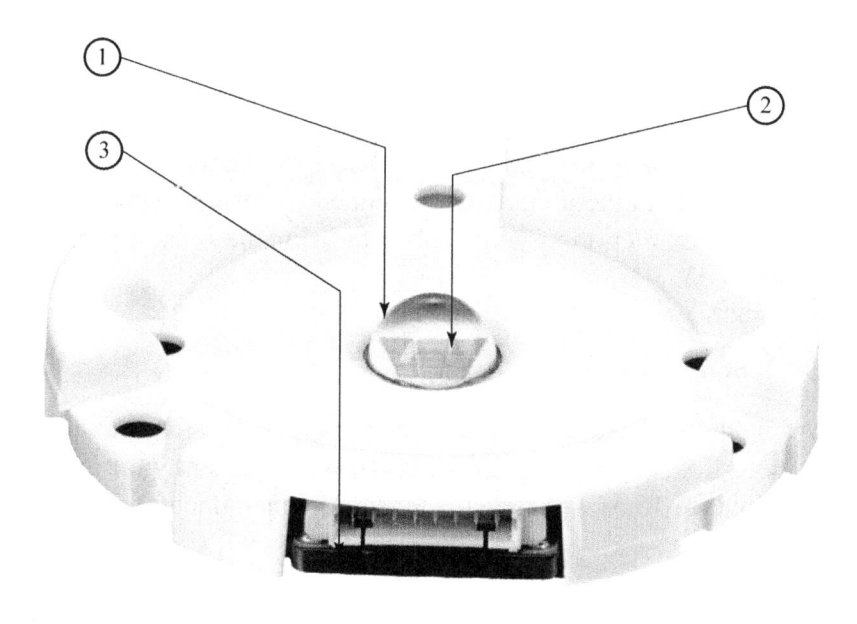

In order to maintain the performance specified by the manufacturer, LEDs need to dissipate heat and to be driven stably.

15.4.2 Heat dissipation

Heat dissipation is probably the most important feature that the fitting must provide to the LED in order for it to work.

The average rated life of the LED is terribly sensitive to heat: just ten degrees out of the recommended specification and the average life halves.

Figure 99: LED cooling: active on the left and passive on the right.

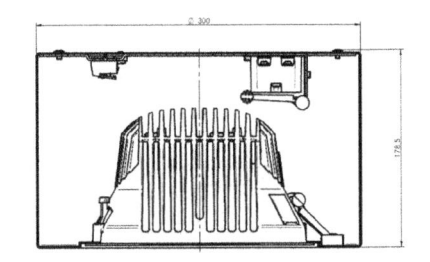

LEDs do not produce significant amounts of heat, however they concentrate it in an area just a few square millimetres wide, and since heat dissipation is a function of the area of the exchanging surface, this makes things rather difficult. To solve this problem (see Figure 99), manufacturers can add finned metal heat sinks, or, an active heat extraction system with a fan (for the most demanding applications).

15.4.3 Stable driver

Current stability makes the LED perform at its best continuously: luminous flux, colour temperature and most importantly the average rated life (see 15.5) will remain close to the manufacturer's specifications.

A high quality LED driver is paramount, both for the performance and for the stability against failure of the LED and driver.

15.5 Average life

The traditional definition of average life does not apply well to LEDs. Solid-State Lighting (SSL) requires a slightly different approach[2]. We can refer to two familiar concepts ("11.2.5 Average rated life (B50)" on page 128): rated life, and lumen maintenance life[3]. If we defined failure as a situation when half the LEDs in a batch emits less than 70% of the initial flux, then the combination $B_{50}L_{70}$ gives us the average rated life. For example an average rated life ($B_{50}L_{70}$) of forty thousand hours means that after forty thousand hours 50% of the samples will have a flux that is 70% of the initial value.

It is important to note that in the average rated life definition there is no mention of the duration or reliability of the driver, which is usually the first to fail.

15.6 Binning

Electronics manufacturing and binning have been going hand-in-hand for a number of years: the process used to produce electronic chips economically can not guarantee that two items coming out of the production line one after the other will share the exact same performance. It is necessary to test every item and group them into bins according to the resulting performance.

The current level of efficient manufacturing technology for LEDs is exactly in the same position: it does not guarantee that two diodes coming out of the production line one after the other will share the exact same characteristics. Luminous flux, colour temperature, colour of light and driving current can be different. This means that LEDs too need to be tested and grouped into bins that share a similar performance.

---

2. (JIAO 2011)

3. (IESNA LM-80 2008)

Binning is, of course, an expensive and cumbersome step in the production of LEDs.

Figure 100: MacAdam ellipses (indicated by the arrow).
Each ellipse contains colours that should not be discernible by the average observer. LEDs from the same bin will belong to a maximum of four of them.

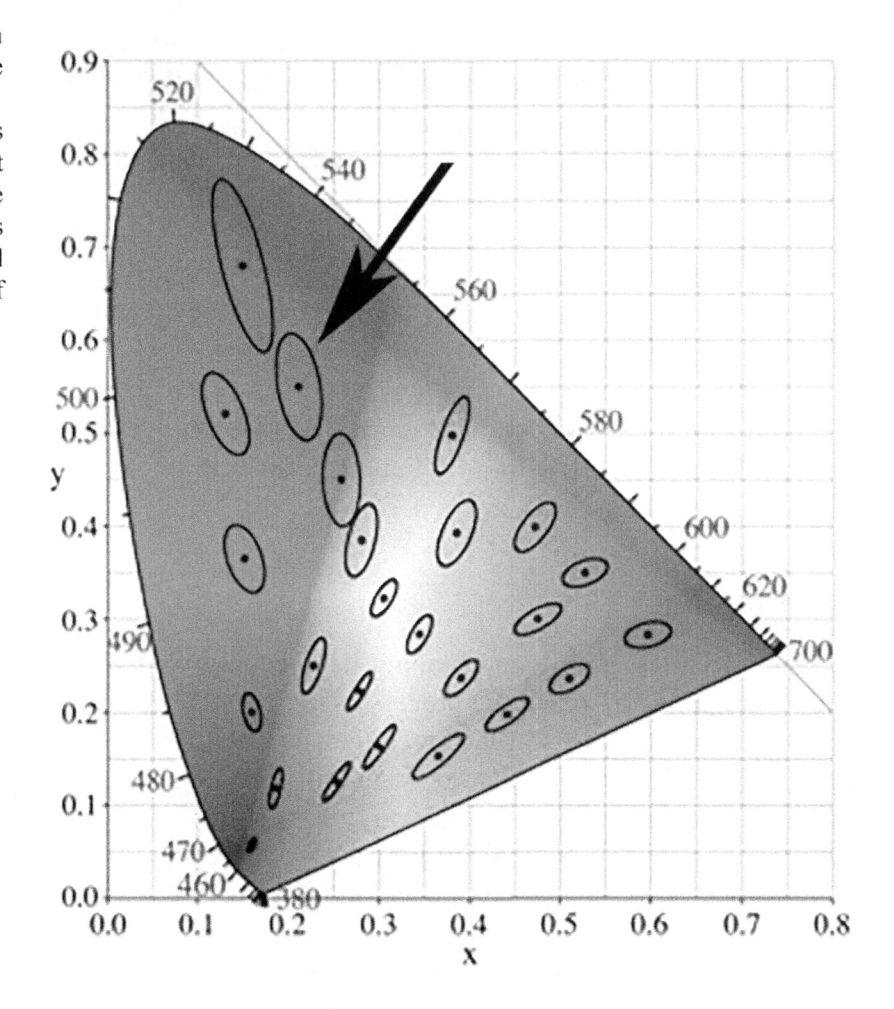

Let's look into colour consistency: Figure 100 shows MacAdam ellipses: each contains colours that the average observer perceives as exactly the same.

Unfortunately, a bin contains LEDs spanning up to four ellipses, i.e. with visibly different colours, so those applications requiring ultra high colour fidelity need an expensive, manual selection after the binning process.

These minute differences are the reason why it is impossible to reproduce a specific colour at a high level of fidelity with a RGB LED: the monochromatic emitters will have slightly different colours and their flux will change with current in a different way.

This of course does not mean that LEDs will not be able to easily produce millions of colours. It's just impossible to precisely select one of them.

15.7 Luminance control

In "9.4.1 Bright lamps" on page 109, I talked about the need to protect the observer from lamps producing excessive luminance. Let's calculate the luminance of a LED and see where it places on "Table 11: Minimum screening angles for high luminance light sources, according to EN12464-1." on page 109.

If we expect an emitter surface smaller than 4 mm^2 and an intensity of about thirty-three candle, the luminance will turn out to be 8.5 million cd/m^2, or 17 times the minimum requirement for the highest screening. This is such an enormous amount of luminance that the LEDs will surely damage visual comfort even if they are not directly in the field of view of the observer. This is also why it is unwise to use naked LEDs unless they are either completely hidden from the observer or covered by a diffusing screen or lens.

15.8 Evaluating grid

15.8.1 Bulb and cap

LEDs reach by far the smallest sizes possible for a light source, allowing new creative ways of lighting. A further advantage is that RGB LED do not take up any more space than monochromatic or white ones. This lets designers come up with creative solutions for changing the mood of an interior.

On the other hand, LEDs are used in retrofit for fluorescent tubes, so the size range goes from the smallest to the largest.

The caps of retrofit lamps are obviously exactly like the ones they replace.

Table 24: Evaluation table for LEDs.

Indicator	Performance
Bulb and cap	very small to large
Colour temperature	warm and cool
Colour rendering	from average to good
Luminous efficacy	high
Average life	uncertain - from 25000 h to 40000 h
Electrical devices	(almost) always

15.8.2 Colour temperature

Both warm and cool colour temperatures are now available.

15.8.3 Colour rendering

The traditional definition of colour rendering index is not applicable to LEDs. In the meantime LEDs have their CRI, which is useful mostly to distinguish between a high quality source and a low quality one.

The first white LEDs had awful colour rendering and very cool light, while the more modern versions have warm light and a much better colour rendering index. The performance is so good that LEDs have started being used in museums.

15.8.4 Luminous efficacy

Solid-state lighting improvements happen rapidly, not only because the technology is not mature, but also because the improvement in electronics is much quicker than what we are used to with traditional lamps.

Luminous efficacy has increased immensely from the first LEDs, and, at the time of writing, it is on level with compact fluorescent lamps.

15.8.5 Average life

Even though there is no standardised definition and, probably because of that, this number has been grossly overstated as a marketing ploy. LEDs have the longest average life of all the light sources: sensible numbers today are in the range of 25,000 hours

15.8.6 Electrical devices

There are one or two cases of LEDs not using a driver, but they are mere curiosities. It is not wrong to expect all LEDs to need a driver.

15.9 OLEDs

Figure 101: Example of OLED application (Philips Living Shapes). The OLED turns on and off according to the movement of the person. When off, these module look like mirrors.

Organic LEDs (Figure 101) use carbon-based semiconductors instead of silicon-based ones for their emitters. The advantage is that they are able to emit wider spectra of light thus improving colour rendering, and that they can be transparent or mirrored when off.

Because of the really high costs of this technology (from 90 € to 350 € per OLED and about 50 € per driver), and the very low luminous efficacy, this type of source is not much more than a prototype. Considering though that the rate of innovation of both the materials and the performance for LED lighting has been phenomenal, I can't but expect them to reach a very wide diffusion in the near future.

15.10 Example of fittings

LEDs are traditionally used in applications where miniaturisation of the source is paramount, such as hidden cove lighting; where maintenance costs are high and where the designer needs coloured light effects, such as colour changing lighting of public monuments and architectural features. They were mainly a task lighting kind of source, and, until recently, their preferred incarnation was the spotlight indoors and the projector outdoors.

Lately they appear in almost any type of light fitting and application, because their efficacy and light quality has improved significantly.

Figure 102: LED luminaires for general lighting.

15.11 Self evaluation

Answer the following questions to test your understanding of the material:

LEDs

1. Describe the characteristics of LEDs as light sources. Are there different types of them?
2. List the characteristics of LED lamps that contribute to the evaluation table.
3. Describe the strengths and weaknesses of LED lamps.
4. Explain what is binning.
5. Compare the cost of LED lamps with equivalent halogens.
6. Describe in detail the average life of LEDs and their colour rendering.
7. Give examples of light fittings that employ LEDs and of fields of application.

Figure 103: Total costs of the different solutions.

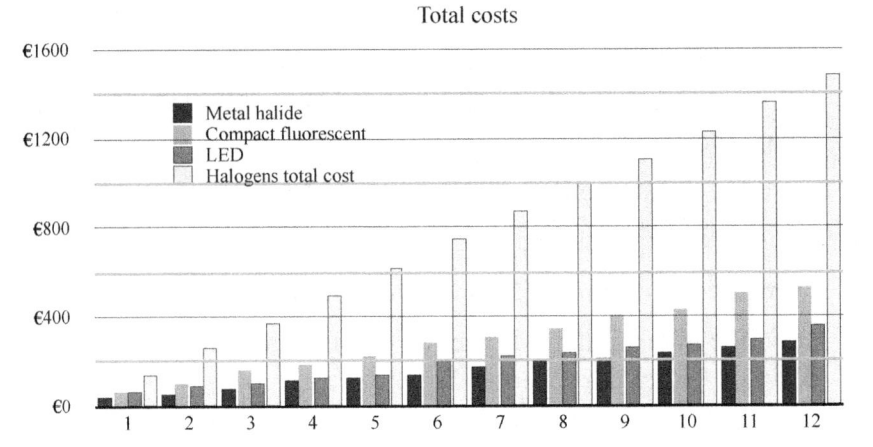

Figure 104: Lamps used in the different solutions.

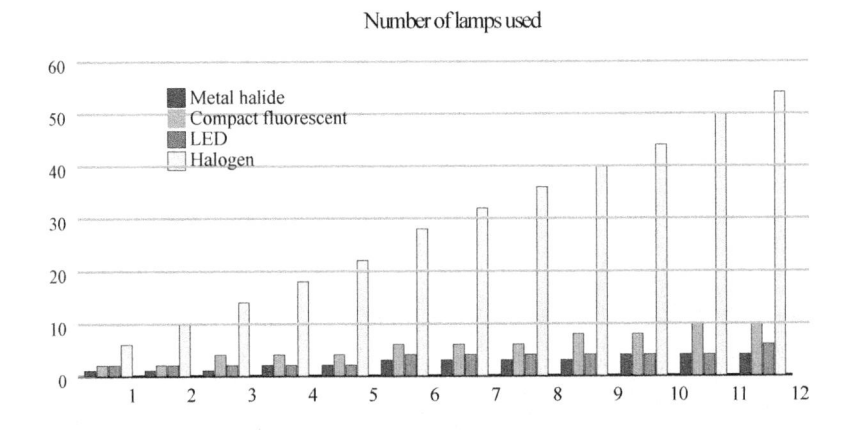

Figure 105: Yearly cost of the different solutions.

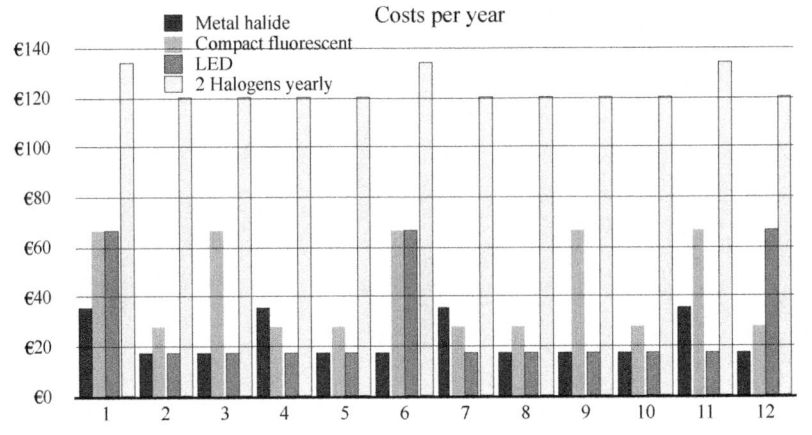

Economic comparison between lamps 16

I want to quickly summarise the results of the economic comparison between lamps that I carried through this part of the book.

The cheaper overall solution (Figure 103) is the metal halide lamp. If the application does not require instant start up, then a metal halide lamp produces very high quality light at an excellent luminous efficacy.

If instant start-up is necessary, then both compact fluorescent and LED lamps provide excellent alternatives, with a slight preference for LED where efficacy is important, and compact fluorescent where light quality is important.

It is very difficult indeed to suggest the use of halogen light source today.

Figure 105 compares costs for the better solutions, in particular it details how expensive it will be, each year, to select each light source.

Finally, it is important to remember that these comparisons will change in time, because the rate of improvement of LED light sources is very quick compared to traditional sources.

Part 3

In this chapter we will use the our newly acquired knowledge to study light fittings, their characteristics, performance and so on.

I shall first define the visual task, then the luminaire in general and finally look into those indicators of performance that can group the different luminaires into sensible sets, instrumental to the design process.

The final objective is to give the reader the ability to answer the following question:

```
Given a visual task and an activity per-
formed on it, what is the best luminaire   (54)
for it and why?
```

Figure 106: A headlamp for a coach (end of XIX century).

Luminaires: general description

17

This chapter deals with the light fitting in general, introducing the two auxiliary concept of visual task and of ingress protection.

17.1 Functions and parts of a luminaire

A lighting fixture is some sort of enclosure that has the historic function of containing the light source and protecting both the source and the exterior world from coming into contact with one another.

The interior must be protected from the exterior for a series of reasons: some lamps have a quartz tube that can be damaged by the oil naturally present in the skin; the reflecting surfaces need to be kept as clean as possible; there may be water vapour, rain or the fixture may be submerged in water.

Figure 107: Reflector and lamp of a street lighting luminaire.

At the same time the environment must be protected from entering into contact with what is inside a luminaire: there is a very real risk of electrocution, lamps usually reach high temperatures and the reflectors can be sharp.

The ancient luminaire in Figure 106 has a glass and metal casing that protects the candle and the environment from each other.

The IP rating (see 17.2) can be used in order to assess the effectiveness of a container to protect from external agents.

Another very important aspect of the luminaire is its optical properties, or the ability to obtain a specific light intensity distribution curve. This determines the luminous behaviour of the fitting and the type of visual task that the luminaire can light well.

Figure 108: Fresnel lens.

The elements that shape the light are called optic components, and they are: reflectors, refractors and diffusers.

1. Reflectors (Figure 107) can be mirror-like or matte; they are usually made of aluminium, brightened and anodized, and they are the main optical component. Their function is to reflect the light coming from the lamp so that it is concentrated in specific

directions and dispersed in others. It is the main generator of the light intensity distribution in the fixture.

2. Refractors are usually made of glass, transparent plastics, acrylic resins, or epoxies (PMMA). They can be passive (just a protection against the environment) or active, i.e. contribute to the shaping of the light. An example of an active refractor is the Fresnel lens (Figure 108) that is usually mounted in front of a spotlight to modify the beam. Fresnel lenses were first used in lighthouses.

Figure 109: Compact fluorescent fitting with diffuser.

3. Diffusers (Figure 109) are made of much the same materials used in refractors, the difference being that they have a satin appearance, because their function is to reduce luminance and remove shadows from the scene. The presence of a diffuser means that the light intensity distribution will never be close to the one required for illuminance uniformity, but it will be Lambertian (like a sphere).

17.2 Ingress protection and the IP rating

In order to certify the actual performance of the enclosure, European regulations use an index called ingress protection rating (IP).

The IP rating is described by the letters "IP" followed by four characters (IPXYAB), with X and Y being numerals and compulsory in every designation, and A and B being not compulsory and not important for us.

Table 25: IP rating: description of the first numeral.

IP first numeral

#	Description	Definition
0	Non protected	-
1	Protected against solids $\varnothing \geq 50$ mm	The sphere $\varnothing = 50$ mm shall not fully penetrate
2	Protected against solids $\varnothing \geq 12.5$ mm	The sphere $\varnothing = 12.5$ mm shall not fully penetrate; the jointed test finger ($\varnothing = 12$ mm, $L = 80$ mm) will be clear from hazardous parts
3	Protected against solids $\varnothing \geq 2.5$ mm	The sphere $\varnothing = 2.5$ mm shall not penetrate.
4	Protected against solids $\varnothing \geq 1$ mm	The sphere $\varnothing = 1$ mm shall not penetrate.
5	Dust protected	Dust can penetrate but not in a quantity to interfere with operation or safety.
6	Dust tight	No dust at pressure of 20 mbar inside.

The first number after "IP" indicates the level of protection against the penetration of solids in the enclosure. Its values and their meaning is shown in Table 25.

The second number indicates the performance against liquids (Table 27).

An IP 20 object, for example, has protection against solids larger than 12.5 mm and no protection against liquids; while an IP45 fixture has protection against solids larger than 1 mm in diameter and protection against water jets.

Luminaires: general description

Some applications require a specific IP rating, such as fixtures that need to operate in high humidity environments or underwater in swimming pools.

Zone	Description	Requirements
0	Inside the tub / the shower basin	min IPX7 12V SELV
1	Over the tub / shower basin up to 2.25m from the ground	min IPX4 (IPX5 if water jets)
2	All around zone 1 60cm wide over wash basins at up to 60 cm from the tap	
3	All the rest	

Table 26: Bathrooms: zones and requirements.

Bathrooms are a good example: they are a special location for electrical and lighting installations because of the high risks of shock due to the proximity to water. The room is divided into zones (Table 26), from 0 to 3, with the smaller numbers closer to water, and for each zone there is a set of requirements to satisfy. For example for fixtures mounted into the tub, the IP rating required is IP67, and they need to be low tension; for fixtures mounted over the tub or closer than 60 cm from it, or over the sink, the requirement is IP44. Zone 1 and 2 share the same lighting requirements.

17.3 Visual task and activities

When a designer studies an interior, one of the steps is the careful evaluation of all the activities that are performed (see "24.2.4 Activities and the Complete Activity Table" on page 305)in order to associate to each a subset of the lighting system. Without going into too much detail, an activity can be something necessary to the function of the interior (such as eating in a restaurant), or something necessary to the success of the interior (such as showing a painting in the restaurant, or any other winning feature of the interior).

Table 27: IP rating: description of the second numeral.

IP second numeral

#	Description	Definition
0	Non protected	
1	Protected vs. vertically falling water drops	No harmful effects from vertically falling drops
2	Protected vs. vertically falling water drops when the enclosure is tilted to up to 15°	No harmful effects from vertically falling drops even when the enclosure is tilted at up to 15°
3	Protected vs. spraying water	Up to 60° on either side of the vertical
4	Protected vs. splashing water	Splashed from any direction
5	Protected vs. water jets	Jets from any direction
6	Protected vs. powerful water jets	Powerful jets from any direction
7	Protected vs. temporary immersion in water	Under standardised conditions
8	Protected vs. continuous immersion in water	Under standardised conditions more severe than numeral 7

Each activity will be performed on a set of locations, called visual tasks (see also "6.1 Visual task" on page 61). It is a simple concept, but it has very important ramifications.

Let's consider an example: the till area in a store is certainly a place where an important activity is performed: the extraction of money from hopefully willing customers. When anyone thinks about a store, it is one of the most important activities that comes to mind, especially to the owner. Any mistake made by the cashier because the place is badly lit is not acceptable.

Luminaires: general description

In the same store, the pictures on the walls may or may not be relevant: if they are family pictures of the owner, they are probably not so important for the success of the store... unless the store happens to be an art gallery, or a photographer's studio.

The importance of activities and visual tasks is clearly recognised by the regulations[1], since they assign to each a set of requirements.

It is not enough to concentrate on the visual tasks: doing so may produce excessive contrasts of luminance. A good lighting system must illuminate the surrounding areas of the visual task as well, producing less illuminance (because the surroundings are not where the action is, after all) the farther away one goes.

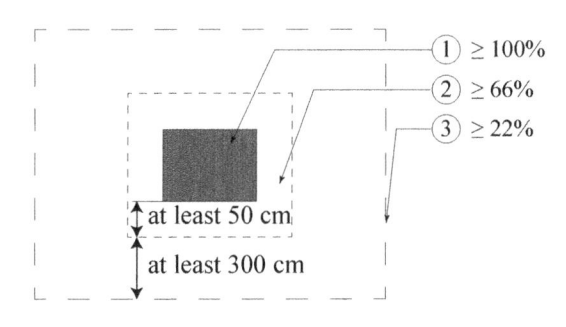

Figure 110: The task area must not be an island of light in a sea of darkness.
1: task area
2: immediate surrounding (width at least 0.5 m) around the task area
3: background area (width at least 3 m from the task area, within the limits of the space)

This common sense idea has been coded by the regulations as well. They define (Figure 110) a "surrounding area" as a band at least 50 cm wide right next to the task area, and require for it (see also Table 28) an illuminance that is about 2/3 of the task area's (so if the task area is 100%, the surrounding area will be over 66%).

Moreover, in particular for work areas without daylight, the regulations also define a "background area" as a band at least 3 m wide, and require for it a maintained illuminance that is at least 1/3 of the illuminance of the surrounding area (so if the surrounding area is over 66% of the task area, the background area will be over 22%).

---

1. (EN 12464-1 2011) and others.

When the illuminance requirement for the task area is low, the surrounding area shares the levels of the task area. Table 28 provides the details.

Table 28: Relation between the illuminance of the task area and that of the surrounding area.

Illuminance on the task area E_{task}	Illuminance on the surrounding area
≥ 750	500
500	300
300	200
200	150
150	E_{task}
100	E_{task}
≥ 50	E_{task}

There are situations[2] in which the designer might want to increase the levels of maintained illuminance over the requirements; these are when:

1. Visual work is critical.
2. Errors are costly to rectify.
3. Accuracy, higher productivity or increased concentration is of great importance.
4. Task details are of an unusually small size or low contrast.
5. The task is undertaken for an unusually long time.
6. The visual capacity of the worker is below normal[3].

What if the size and or locations of the task areas are unknown to the designer? The best solution is to consider the whole area as task area, and light it accordingly. The alternative is to design a flexible

---

2. (EN 12464-1 2011)

3. Visually impaired people have different requirements.

system that can be quickly adapted or moved to provide the necessary illuminance when the locations become known.

17.4 Self evaluation

Answer the following questions to test your understanding of the material:

1. Define "visual task", describe the regulations that deal with it and explain how the lighting system should work with it.
2. Describe the historic and technical function of the light fitting.
3. Explain what is the Ingress Protection rating.
4. Explain how a luminaire can produce its light intensity distribution.

Figure 111: Sample room for luminaire selection by flux, from DIALux.

Figure 112: Sample room for luminaire selection by flux - plan view from DIALux.

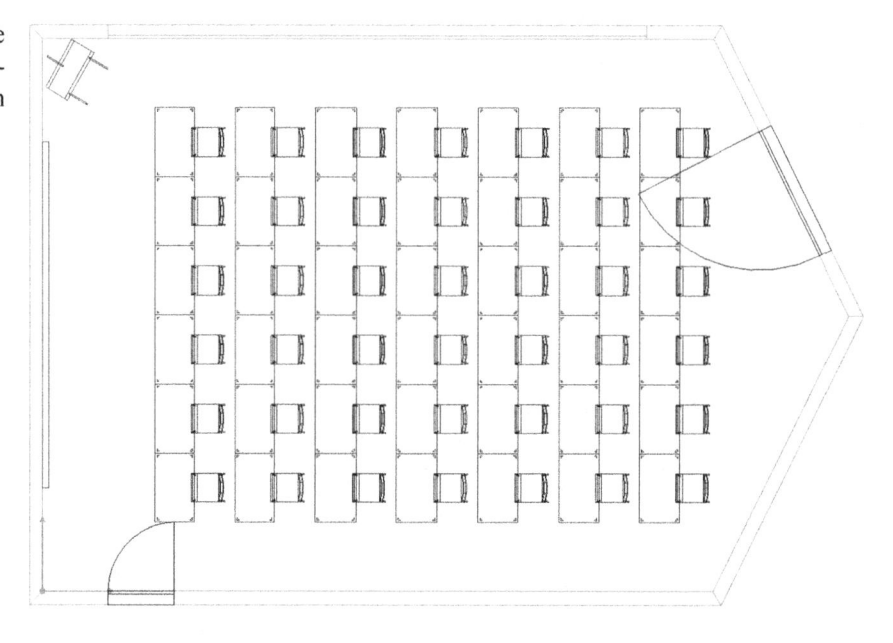

Selecting by flux: CIE classification 18

The CIE (international commission for lighting) proposes a series of classes according to the amount of luminous flux that is emitted over or under the horizontal plane.

When a designer imagines his interior, he will have an idea of what kind of light he wants: will there be harsh shadows or diffusion, will the ceiling be lit or dark, will the walls be bright or not, and so on. Some of these effects are shown in "23 The concept" on page 285. Using this particular criterion permits to select only luminaires that send light where it is wanted.

Table 29 shows the specifications for each class.

Class	Percentage of flux going upwards	Percentage of flux going downwards
Direct	0 - 10	90 - 100
Semi-direct	10 - 40	60 - 90
General-diffused	40 - 60	40 - 60
Direct-indirect	40 - 60	40 - 60
Semi-indirect	60 - 90	10 - 40
Indirect	90 - 100	0 - 10

Table 29: Specifications for each CIE class.

18.1 Sample room

I have selected a sample interior (Figure 111 and Figure 112) where I can test the different types of fixtures, and compare the results. It is the same classroom/conference room I will use in "Part 5" on page 315 to talk about the lighting calculation software. It is 12 metres by eight, with five rows of desks for the attendees, each with six chairs, a large window on the north wall, two doors and a whiteboard on the west wall.

18.2 Direct

Figure 113 shows on the left a sketch of the direct fixture. It is a box with an opening for the light on the bottom, with or without diffusers.

These light fittings emit practically all their flux downwards. They can be recessed, and thus completely hidden from the viewer and, if equipped with a diffuser or a screen, they do not break the lines of the interior.

Figure 113: Direct fitting. From left to right: a sketch of the luminaire and its light intensity distribution.

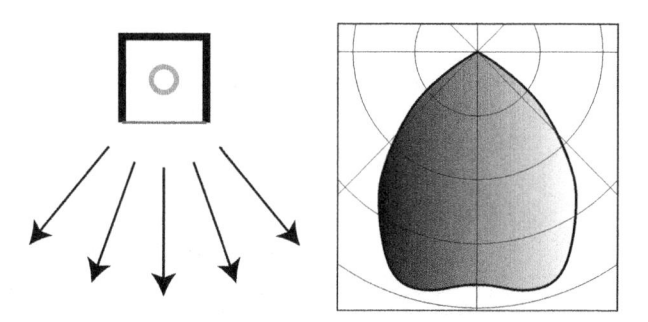

The most common example that comes to mind is the downlight (see 22.6.1 on page 271).

Figure 114: Direct example. From left to right: a picture of a ceiling mounted fitting and its photometric curve.

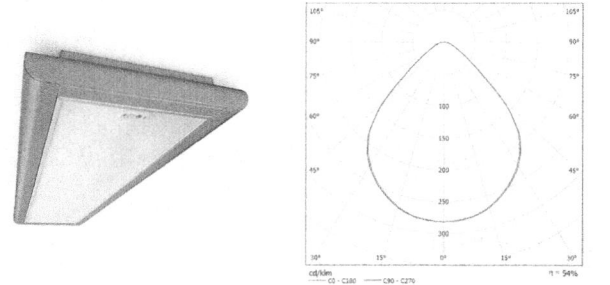

Another example that might come to mind is the table lamp. It was a major player in office lighting before the advent of high quality fluorescents because it was traditionally the only way to generate high illuminance on the work plane with low powered light sources.

Today there are fewer reasons to employ them because of the inevitable worsening of visual comfort they cause, as they increase the luminance of the foreground and do nothing for the background, often even failing to illuminate the whole task.

On the right side of Figure 113 there is a very approximate sketch of the light intensity distribution curve.

Figure 114 shows an example of luminaire (a fluorescent one with a diffuser), and on the right its light intensity distribution curve.

Unfortunately, the diffuser produces a light intensity distribution that is incapable of providing uniformity.

Figure 115: Sample interior lit by direct fittings. Notice how both the ceiling and the top of the walls are darker.

The advantage of this category of fixtures is that they are less sensitive to the reflectance of the room, as the tasks receive most of their light directly, so the interior designer is free to select low reflectance colours and materials for walls and ceiling without lowering the efficiency of the system too much. The sample room (Figure 115) with white walls needs six luminaires to be compliant, but if the walls were black, the room would require just two more light fittings, quite a minor increment for such a reduction in reflectance.

Unfortunately, emitting luminous flux downwards only leaves the ceiling darker than the surrounding walls (see also Figure 213 on page 291), thus creating a weird feeling in the room.

18.3 Indirect

These light fittings emit almost all their flux upwards. Figure 116 shows a sketch of the indirect fixture and of its light intensity distribution curve.

Figure 116: Indirect fitting. From left to right: a sketch of the luminaire and its light intensity distribution.

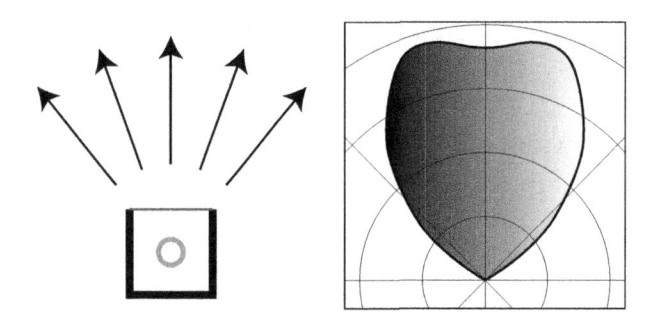

The most common type is the wall mounted luminaire (Figure 117). The advantage over fully diffused fittings is that direct glare is impossible because no intensities are emitted in the direction of the eyes of the observers. Still mounting the luminaires too low on the wall, or on very reflective surfaces can definitely cause glare (Figure 118).

Figure 117: Indirect example. From left to right: a picture of a fitting and its photometric curve.

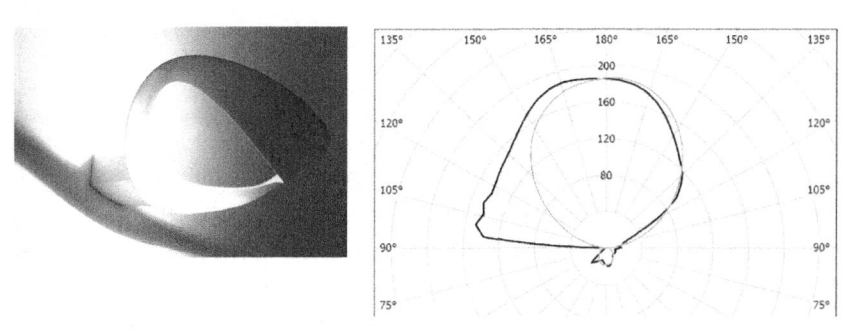

The visual result (Figure 118) is the opposite of the previous one: the ceiling is brighter than the walls, this gives a more natural feeling, as we are used to the sky being brighter than the ground outdoors.

With an indirect fixture the surface finishing, colour and material of walls, ceilings, curtains and furniture become a major factor in the final result, for two main reasons:

1. Reflectance

The quantity of luminous flux that reaches the work plane, via a ceiling or a wall, is determined by the colour, material and finishing of the wall or ceiling. If it is a rough black wall for example, the quantity of luminous flux reflected off the object can be just about 10% of the incident flux. If on the other hand the walls were glossy and white, the reflected flux could reach 70-80%.

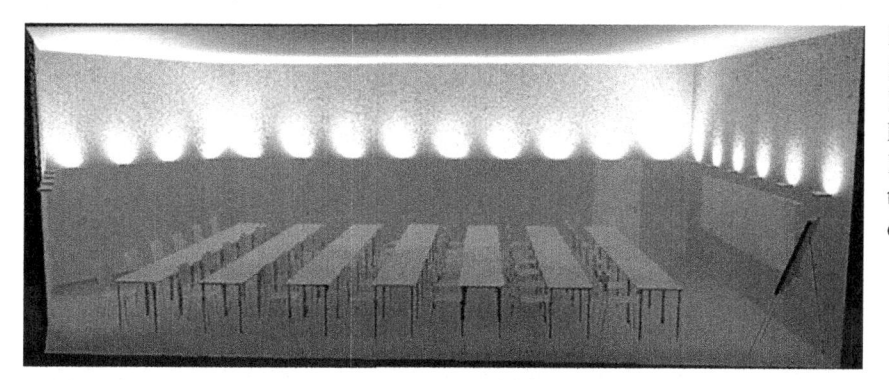

Figure 118: Sample interior lit by indirect light fittings. Notice the high luminance pools of light on the walls and the darker centre of the ceiling.

2. Colour bleeding

The problem with using e.g. coloured walls is that the reflected light will have the same colour of the wall. This means that every object illuminated by this light will have a fake colour. If the extension of the coloured surface is large relative to the space, this may mean that the colour of every object in the interior is unrealistic. For example: if the walls are red, every white object will look red, and every blue one will appear black (see "Colour, colour temperature and colour rendering" on page 13).

This is a big problem where colour rendering is even minimally important, and a disaster in places like restaurants where the colour of the food is one of the main points that make the success of the interior.

Indirect luminaires with a less than perfect light intensity distribution curve, i.e. the vast majority of them, also may produce bright spots on the walls, that can cause excessive contrasts of luminance.

This high luminance area will end up right in the middle of the field of view, so it can seriously damage visual comfort.

It is important to verify the luminance contrast on the wall and find out whether we have glare or not. This can be done with some approximation via the lighting calculation programme (see 21.7.3.1 on page 466), which will produce something like Figure 119, where the maximum luminance is around 300 cd/m^2 and the minimum around 30 cd/m^2. If we then calculate the luminance contrast we obtain 9. This result is much larger than 5, so there will be glare.

Figure 119: Luminance false-colour view of the interior lit by indirect fittings. White is 300 cd/m^2, black is 30.

In a very large room there is also a problem of uniformity: the outer part of the room, closer to the walls, will receive much more luminous flux than the centre, and illuminance will then not be very uniform. In the sample interior, using white walls and ceiling, there is about a 20% decrease in illuminance from the sides of the room to the centre. Moreover, if one of the walls has a row of windows, or any other architectural impediment, no wall-mounted fixtures can be used, and the area receives even less luminous flux. In our example this means about half the illuminance of the area in front of the wall opposite.

The most common use of these luminaires is providing diffused lighting (see "23 The concept" on page 285).

18.4 Semi-direct

In order to avoid the pitfalls of both direct (oppressive feeling generally caused by dark ceilings) and indirect fixtures (low efficacy due to indirect lighting), the designer can use this type of luminaire.

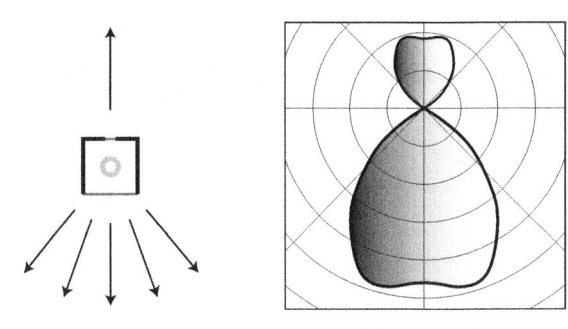

Figure 120: Semi-direct fitting. From left to right: a sketch of the luminaire and its light intensity distribution.

Figure 120 shows a sketch of the semi-direct fixture and of its light intensity distribution curve: it emits almost all its flux downwards (Table 29).

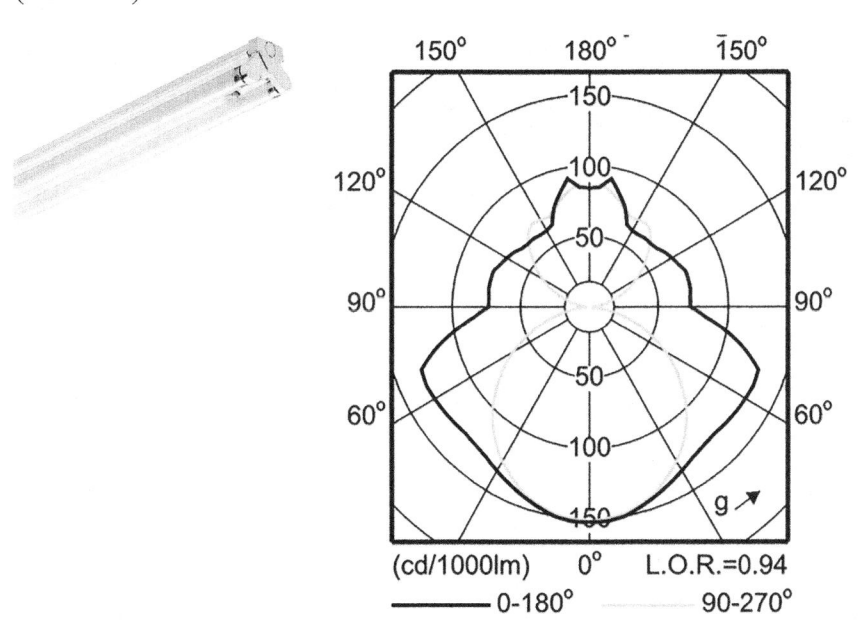

Figure 121: Semi-direct example. From left to right: a picture of a fitting and its photometric curve..

It is an attempt to unite in a single luminaire the functions of both general lighting and accent lighting: the fitting in Figure 121 illuminates the task and the ceiling and walls at the same time, with a strong predominance on the task.

The suspended versions are usually installed in offices, where they may form a line of fixtures that follows the desks in open spaces, or just hover over a single desk.

Being mostly direct they tend to be more efficient and less dependent on the colour and reflectance of the walls (Figure 122).

Figure 122: Sample interior lit by semi-direct light fittings.
Notice the high luminance pools of light on the ceiling, and a darker band on the top part of the walls.

It is important to select the correct mounting height for these fixtures. Install them too close to the ceiling and they will generate long blobs of very high luminance. Install them too low and the luminaires will light just a small area of the floor, while at the same time risk damaging visual comfort.

18.5 Semi-indirect

In this case, the idea is to send the largest part of the luminous flux towards the ceiling to be reflected back on the task area, leaving a small part of it to directly illuminate the task, which is usually a desk (Figure 123).

This light fitting is generally suspended or wall mounted.

Selecting by flux: CIE classification

It is another attempt to have the benefits of both general lighting and accent lighting, this time with the emphasis on general lighting. Ideally the light fitting should have a very wide indirect light intensity distribution (Figure 124), so that it can spread its flux on as wide an area as possible.

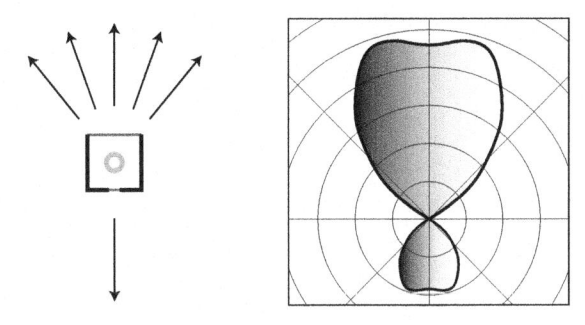

Figure 123: Semi-indirect fitting. From left to right: a sketch of the luminaire and its light intensity distribution.

It is evident from Figure 125 that the bands of high luminance in the ceilings can be detrimental to the visual comfort of the interior. Still, because ceilings do not glare on users so much, this is not a crippling issue.

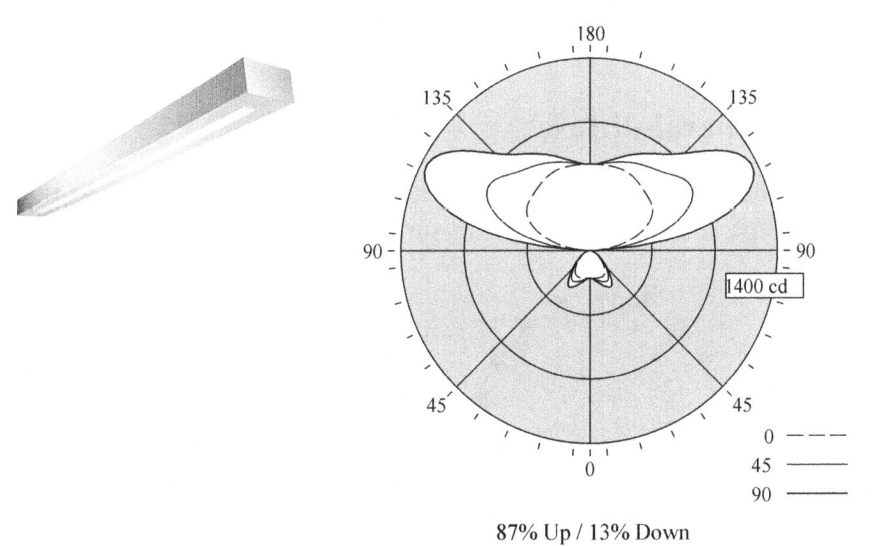

Figure 124: Semi-indirect example. From left to right: a picture of a fitting and its light intensity distribution.

87% Up / 13% Down

In order to reduce this effect it is a good idea to make sure the suspending wires that connect the fixtures to the ceiling are long enough, so that the area illuminated by the indirect luminous flux is large and the luminance contrast is reduced. A wider beamed indirect light intensity distribution obtains the same result.

Figure 125: Sample interior lit by semi-indirect light fittings. Notice the high luminance pools of light on the ceiling.

Luckily, it is much harder for a ceiling to glare down, because its surface is along the direction of view of the observer, so the intensities coming towards his or her eyes are usually small.

18.6 General-diffuse

The light intensity distribution is Lambertian (Figure 126), so not capable of uniformity on any flat surface.

Figure 126: General-diffuse fitting. From left to right: a sketch of the luminaire and its light intensity distribution.

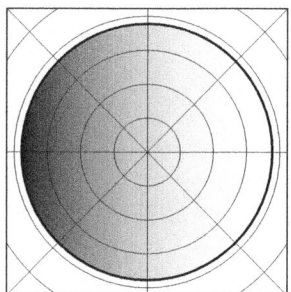

The luminaire is usually entirely surrounded by a diffusing screen (Figure 127) that produces the light intensity distribution.

Selecting by flux: CIE classification

The mounting is commonly either free-standing or suspended.

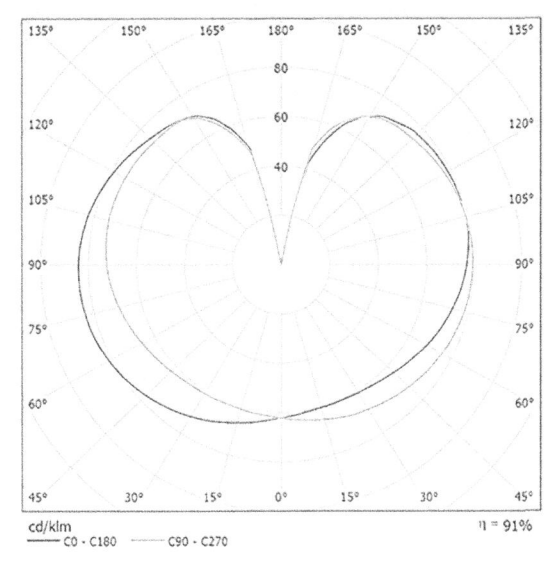

Figure 127: General-diffuse example. From left to right: a sketch of the luminaire and its light intensity distribution.

An interior lit with these fittings only, will appear flat, depth perception will be difficult and there will be no harsh shadows (see Figure 128) because of the excess of diffused lighting.

Figure 128: Sample interior lit by general-diffuse light fittings. Notice the flatness of the atmosphere.

The performance of these luminaires is not very good:

1. The R_{LO} of these fixtures is not excellent, going from all right to abysmal, because the diffuser can't be too transparent if it has to spread the luminance emitted by the lamp and reduce luminance contrasts;
2. The light intensity distribution curve is a circle, thus it is incapable of giving uniform illuminance anywhere, no matter the positioning and the shape of the task (apart from one absolutely surreal case[1]);

The main use for this type of luminaires is providing diffused light in the interior (see "23 The concept" on page 285) and at the same time illuminating the ceiling and the walls. A system comprised of direct recessed diffuse fittings and indirect asymmetrics should be used in larger spaces to obtain the same diffusion..

General diffuse fixtures are by far the most commonly used in residential interiors. The ones with the highest style content are fixtures designed to be seen, not to let people see, and they are usually placed in a position of prominence. It is important that their shape is chosen in accordance to the rest of the interior, and that they have some sort of creative or innovative look to show off.

It may even be wise to find other, better performing luminaires to illuminate the general diffuse light fitting and help it stand out in the interior, as if it was a hanging sculpture. In this case the fitting itself becomes a task.

18.7 Direct-indirect

From a performance perspective these fixtures can be considered as an upgrade of the general diffuser.

The attempt here (Figure 129) is to control the luminous flux emitted closer to the horizontal. That is the direction where a high luminance contrast can be detrimental to the observer, and it is usually a direction free of tasks (see Figure 131).

---

1. A spherical room...

Selecting by flux: CIE classification 209

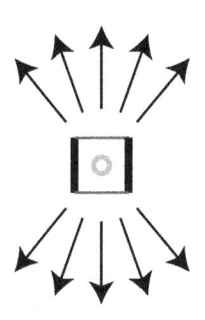

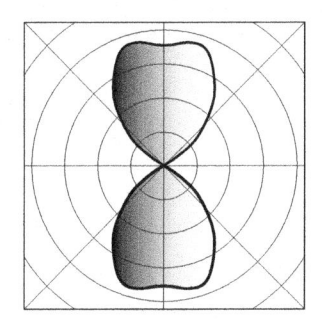

Figure 129: Direct-indirect fitting. From left to right: a sketch of the luminaire and its light intensity distribution.

This means that the fixture (Figure 130) can be placed in the centre of the observer's visual field, to spread its luminous flux in as wide an area of the ceiling and of the floor/task as possible.

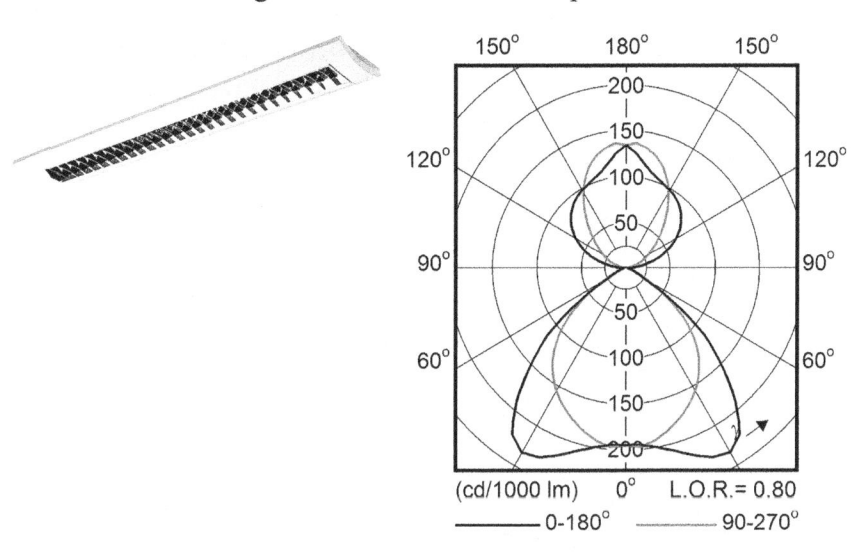

Figure 130: Direct-indirect example. From left to right: a sketch of the luminaire and its light intensity distribution.

Wall mounted fixtures are a little different because in order to illuminate a wide area the fixture must have a very strong asymmetry while being as close as possible to the ideal photometric curve, and that is not so easy to find.

An interior lit by this type of luminaire will have both the ceiling and the task area relatively bright, and the walls darker.

The sample interior (Figure 132) has the ceiling well-lit, but it also has harsh illuminance irregularity on the task, and the whiteboard in the dark.

Figure 131: Sketch of the reduced risk of glare for a direct indirect light fitting suspended in the centre of the field of view.

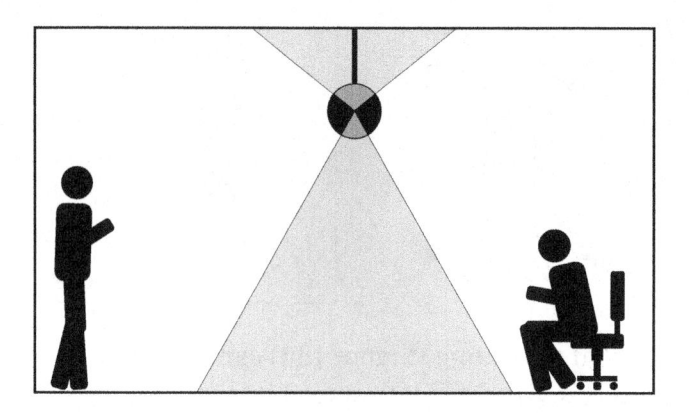

Solving the whiteboard just takes a few luminaires dedicated to it. The irregularities on the desks require more though. The light intensity distribution for this luminaire is too narrow downwards for tasks so wide, and this is what causes the problem. The only way to solve it is to select a fixture with a larger beam spread downwards.

Figure 132: Sample interior lit by direct-indirect light fittings. Notice the high luminance pools of light on the tasks and the bright ceiling.

18.8 Self evaluation

Answer the following questions to test your understanding of the material:

For each of the different CIE types of luminaire:

1. Describe their characteristics.
2. Give examples of luminaires.
3. Give examples of applications.

4. Explain the effect in the interior.
5. Describe their strengths and weaknesses.
6. How can a luminaire be a task? Give examples.
7. What types of luminaires would you employ to:
    7.1. Provide diffused lighting.
    7.2. Illuminate the ceiling and floor, leaving the walls less lit.
    7.3. Illuminate the horizontal visual tasks and leave the rest of the surfaces dimmer.

Figure 133: Sample interior to show the results of different types of fittings.

Figure 134: Sample room - plan view.

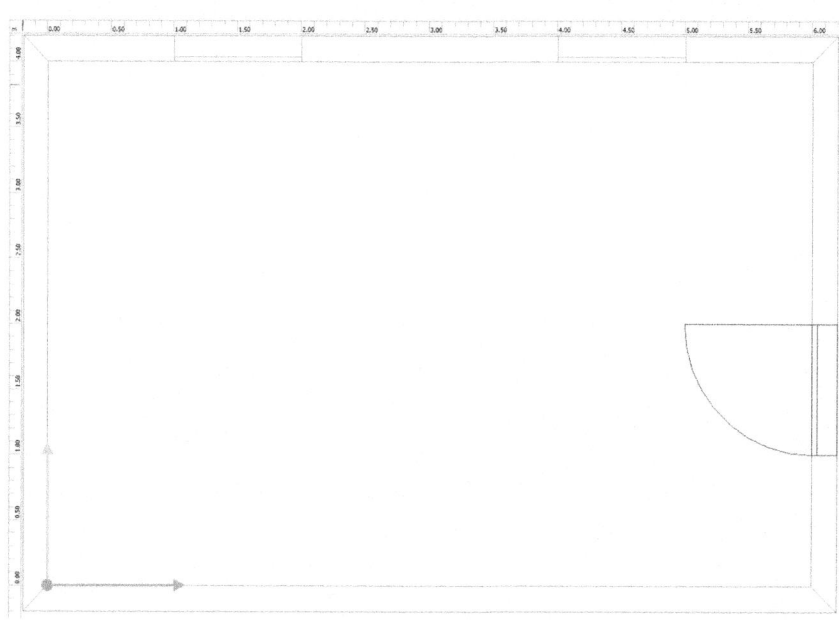

Selecting by LID: Axially symmetric 19

The designer can have a complete picture of what the fixture does, and for what tasks it performs well by evaluating both the photometric curve and the light output ratio. I've talked about LOR starting from "4.3 Light output ratio (RLO)" on page 42, it is now time to deal with the light intensity distribution.

The four types of distributions are:

1. Axially symmetric.
2. Symmetric about two planes.
3. Asymmetric (actually symmetric about one plane)
4. Fully asymmetric.

Before looking into each, we need to define a sample interior.

19.1 Sample room

The interior (Figure 133) is a 6m by 4m room, 3m high. The maintenance factor is 0.8, because it is a clean room. The walls, ceiling and floors are very dark, so that we can observe the effect of the light that comes directly from the fixture, reducing the influence of reflection. That is why reflectance is 12%, 17% and 18% respectively for the floor, the walls and the ceiling.

19.2 Axially symmetric luminaires

These fixtures have the same photometric curve for every C-plane: they emit light the same way all around the vertical at $\gamma = 0°$ (Figure 135). The shape of the light distribution looks like a cake: a slice will look the same no matter from what part of the cake it comes. The light intensity distribution is a solid of revolution around the axis $\gamma = 0°$.

Figure 135: Photometric solid for axially symmetric luminaire. Top view on the left, and side view on the right. It is a solid of revolution.

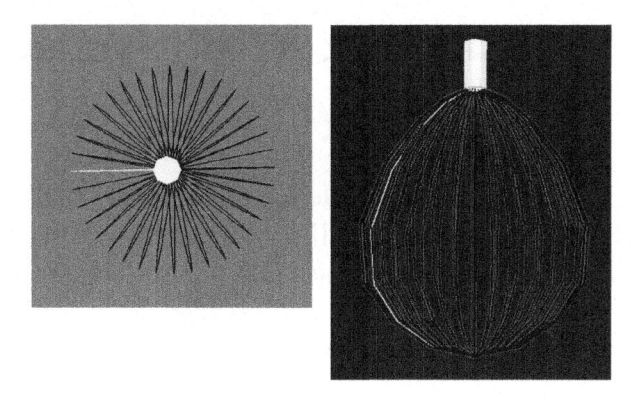

The effect on the floor of a fixture placed horizontally on the ceiling will be a lit circle.

The shape of the fixture is usually conical or cylindrical. Spotlights (Figure 136) are one of the more numerous members of this type.

Figure 136: Example of axially symmetric light fitting, a spotlight, and its light distribution curve.

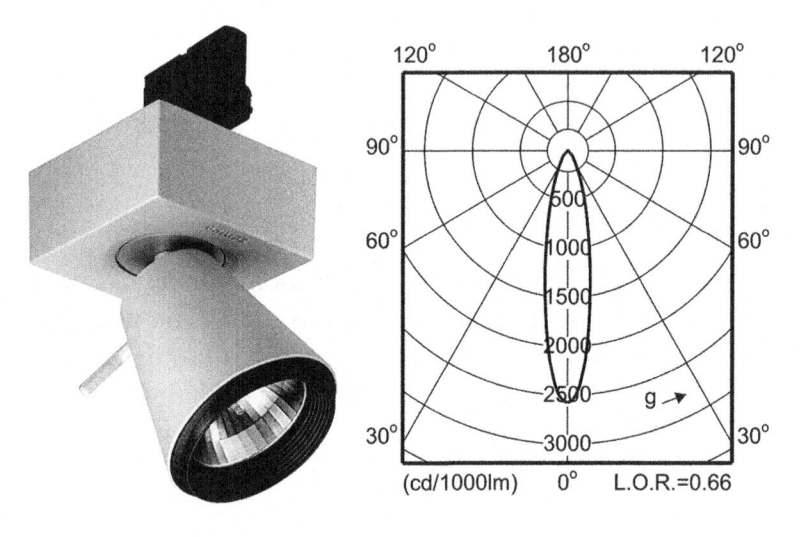

It is useful to further differentiate these luminaires by their beam spread.

19.3 Axially symmetric narrow beam

A picture of this luminaire and its light intensity distribution is in Figure 136. These fixtures can be placed far away from the task thanks to the narrowness of the beam, so as not to impede the use of the task.

Their ideal application is on tasks that are small at least in one direction (see "34.2 Possible improvements" on page 382) compared to the distance to the luminaire.

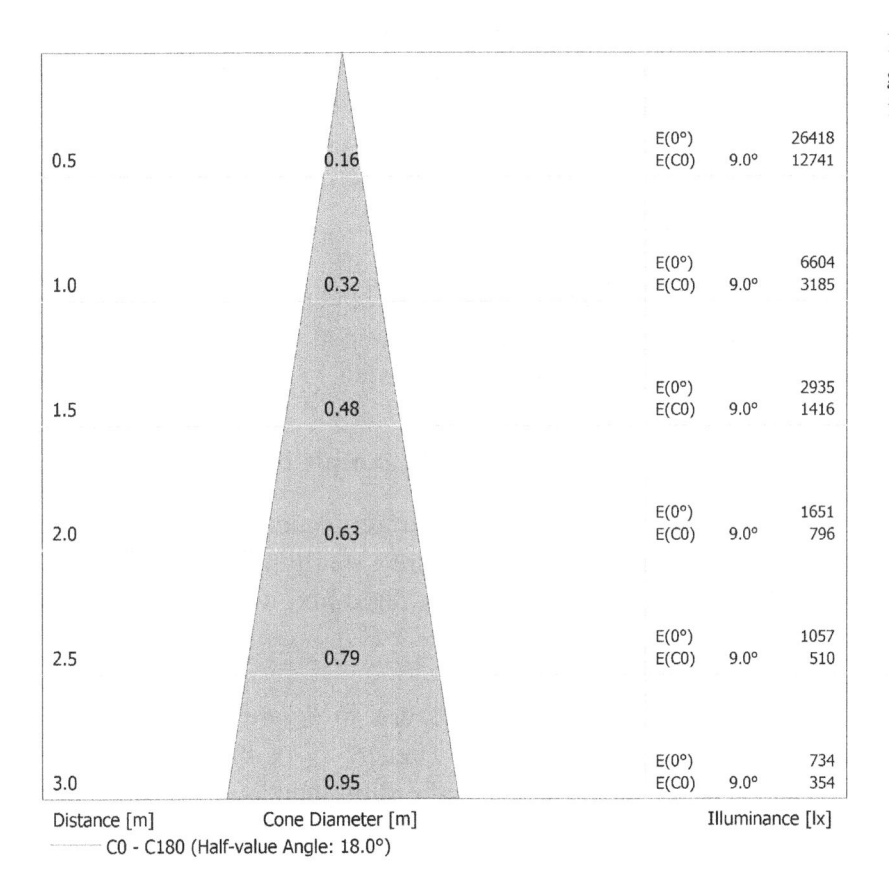

Figure 137: Cone diagram for the luminaire in Figure 136.

Distance [m]	Cone Diameter [m]	Illuminance [lx]		
0.5	0.16	E(0°) E(C0)	9.0°	26418 12741
1.0	0.32	E(0°) E(C0)	9.0°	6604 3185
1.5	0.48	E(0°) E(C0)	9.0°	2935 1416
2.0	0.63	E(0°) E(C0)	9.0°	1651 796
2.5	0.79	E(0°) E(C0)	9.0°	1057 510
3.0	0.95	E(0°) E(C0)	9.0°	734 354

C0 - C180 (Half-value Angle: 18.0°)

The designer can refer to a cone diagram, such as the one calculated by DIALux and shown in Figure 137, to find out whether the luminaire is right for the task. A single luminaire will light a larger task with lower illuminance as the mounting distance increases. For example, the luminaire will create a pool of light 32 cm wide with an average illuminance of 3185 lx if the task is at 1 m distance from the fitting, or produce an average of 510 lx in a circle 79 cm wide if the task is at 2.5 m.

If the fixture isn't perpendicular to the plane of the task, the high luminance area will become an ellipse, and both the average and the maximum illuminance will be smaller[1].

These fittings cast a harsh shadow, and this must be taken into account when designing the mounting positions. The advantage is that this kind of directional lighting increases the sense of perspective; the weakness is that shadows can confuse the viewer and obscure parts of the task. For more details of this, see "23.3 Sharpness and flow" on page 292.

It is also important to consider the flux of people moving in the room when employing directional lighting, to avoid glaring on users because of fittings badly positioned or aimed.

19.3.1 Application: single fitting in sample interior

The result in the room is a very high luminance small circle in the middle of the floor, as shown in Figure 138. Illuminance on the floor right under the fixture soars to over 5000 lux, while the average is just over 70 lux.

Let's look at the isolines for the fixture, in Figure 137: the average must not lead anyone astray: it is a decent 77 lx, but the uniformity is plain horrible: in just a meter the illuminance goes from under 50 lx to over 5000 lx: and almost all the room is under 50 lx. This is a fixture for highlighting, not for general lighting, and the maximum and minimum values confirm this: 5155 lx and 0.87 lx respectively.

---

1. The average will be smaller because the same luminous flux will be spread over a larger area; the maximum because the angle θ between the direction of the luminous intensity and the normal to the task surface will increase.

Selecting by LID: Axially symmetric   217

Figure 138: Render of the sample interior with axially symmetric narrow beam luminaire. Note the narrowness of the pool of light and haze of reflected light on the ceiling.

Figure 139 shows the isolux curves on the floor.

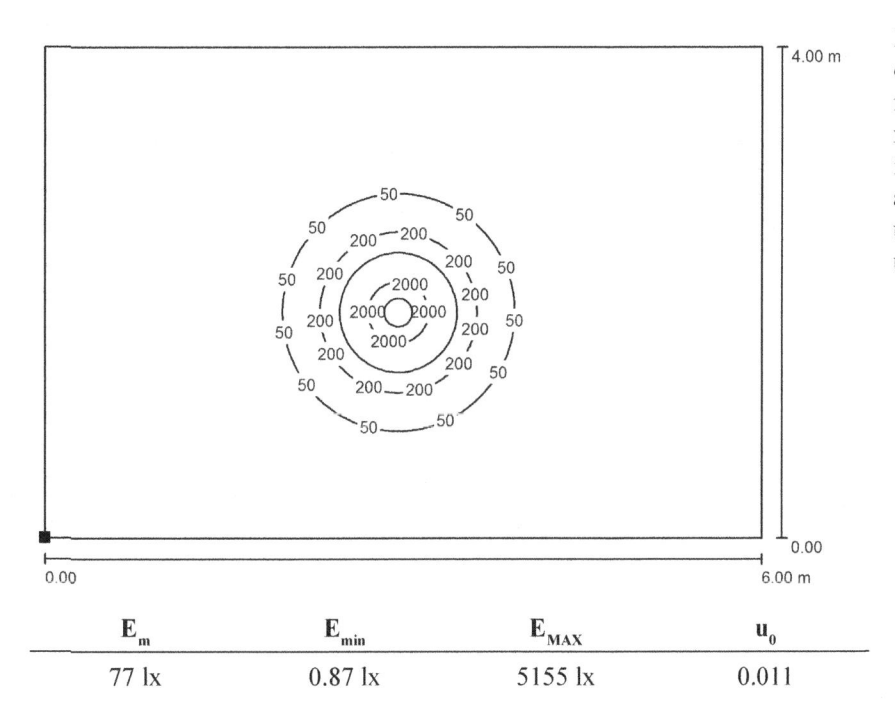

Figure 139: Isolux curves for the axial symmetric narrow beam luminaire in the sample interior. Under the diagram, the table of values. Notice the very low uniformity.

E_m	E_{min}	E_{MAX}	u_0
77 lx	0.87 lx	5155 lx	0.011

19.3.2 Application: sample interior with realistic lighting

In a real life installation we would never employ this type of fitting to illuminate the floor of a room: the result would be a grid of very bright, unconnected areas surrounded by darkness (see Figure 140 and Figure 141).

Figure 140: Awful, but sadly not so uncommon, application of a narrow beam spotlight.

A defining characteristic of these luminaires is their flexibility. Changing the positioning allows the designer to select what kind of illuminance to provide.

19.3.3 Application: horizontal illuminance

If the task requires horizontal illuminance and it is small in size, then this luminaire can be placed high over the task, in a position that avoids glaring and minimizes shadows cast by the viewer. If the task is larger, contained into a glass case or if the designer expects the user to bend over the task to inspect it, then a miniaturised LED lighting solution inside the glass case would avoid the problem of the observer obscuring the task.

Wall grazing is a significantly different way of providing horizontal illuminance using a special narrow beam luminaire. It is also a way to overemphasise the surface roughness of the wall. It consists of a series of narrow beam sources (commonly LEDs), usually hidden in a niche that runs next to the wall. The fittings can be slightly rotated towards the wall to avoid creating a bright line on the floor right next to the wall.

Figure 141: Narrow beam for general lighting.

Wall grazing fittings (22.6.7 on page 277) are shaped as narrow and long boxes, and contain the set of sources needed for this effect.

19.3.4 Application: vertical illuminance

The idea is to illuminate a vertical task, and we will consider one example lit in two ways.

If the task is a flat painting hanging from a wall, the usual solution is a single spot, out of the way, lighting the task without being obscured by the viewers.

The single spot must be mounted high up, so that it doesn't dazzle anyone; but not so high that the shadow of the frame obscures the canvas.

Figure 142 shows the plan view, with the single spotlight and the picture on the north wall. Figure 143 contains a visual rendering of the result.

Figure 142: Painting lit by a single spotlight. Plan view of the system.
1: task
2: single fitting on the ceiling.

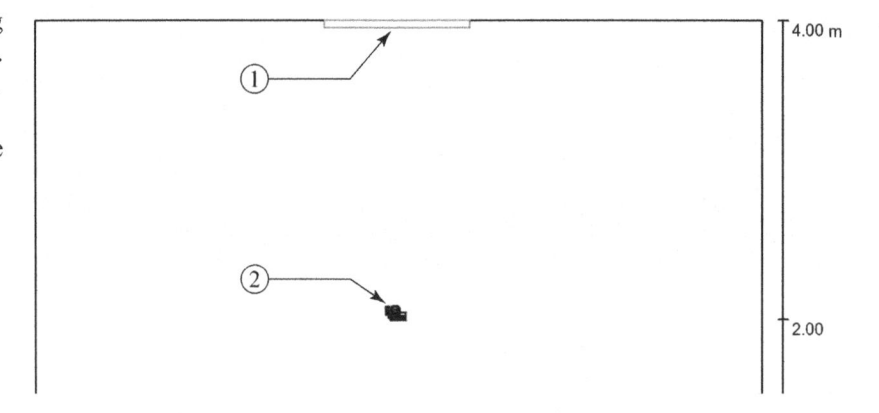

4.00 m

2.00

Notice how the light coming from the spot creates an elliptical pool of light that highlights the painting and illuminates it completely. The bright half ellipse on the wall can give more presence to the picture.

Figure 143: Visual rendering of the picture illuminated by a single spotlight.

Figure 144 shows the isolux on the wall and on the painting. The former has an area where illuminance goes to zero because the part

Selecting by LID: Axially symmetric

of the wall behind the painting is in shadow; the latter shows an average of about 47 lx.

It is important to verify that the frame of the picture does not cast shadows that obscure the picture itself.

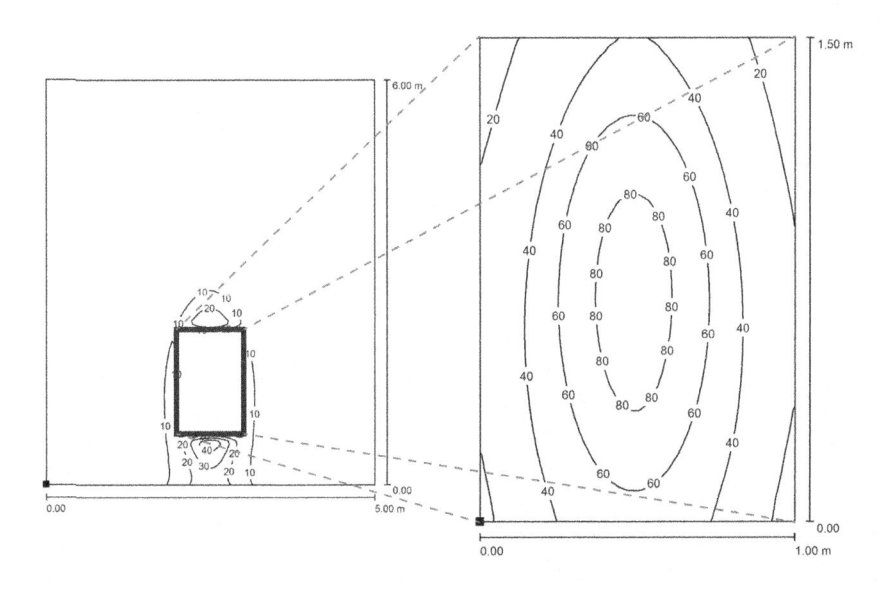

Figure 144: Isolux on the wall (left) and on the picture itself (right) for the single spotlight illuminating the painting.

If the first solution is not feasible, the alternative is placing two spots off to the side (Figure 145) so that even if one is covered, the other will still light the painting.

We can expect the bright spots on the task to be ellipses.

The two spotlights are aimed at two different points: one on top of the other (Figure 147), so that the two bright spots will fill the whole task.

Figure 145 shows the visual result: spotlights only concentrate their flux on the task, so there will be some luminance contrast.

Figure 145: Painting lit by two spotlights. Plan view of the system.
1: task
2: two light fittings on the ceiling.

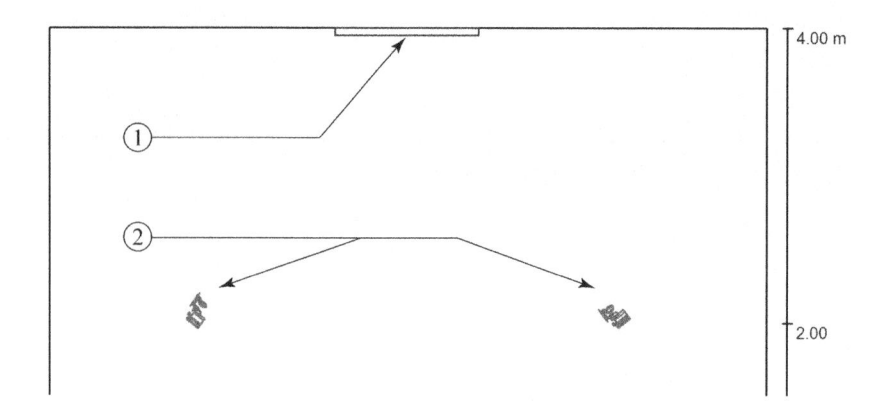

The average illuminance on the task (Figure 147) is almost 60 lx, and the uniformity is not bad ($u_0 = 0.434$). The surrounding wall is of course darker and not so uniform.

Figure 146: Painting lit by two spotlights. Visual render of the effect.

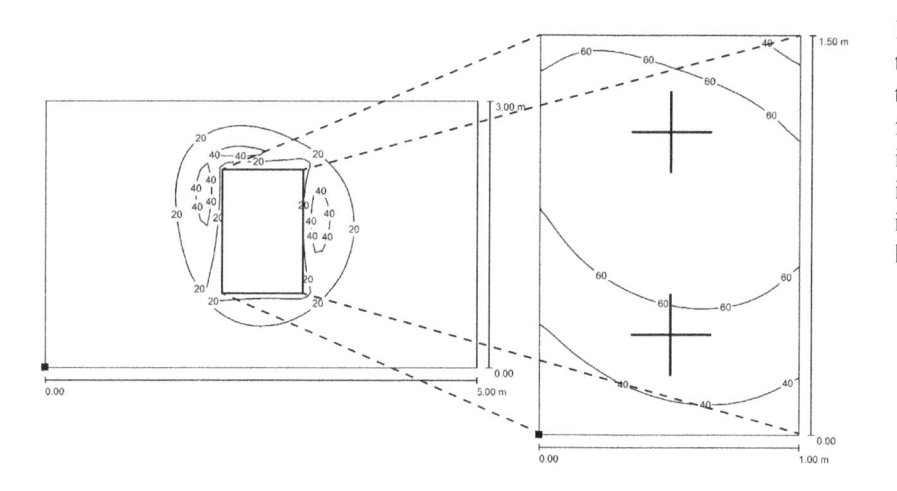

Figure 147: Isolux on the wall (left) and on the picture itself (right) for the two-spot system illuminating the painting. The two red crosses indicate where the spotlights are aimed.

19.3.5 Application: semi-cylindrical illuminance

If, for example, the task is a medieval suit of armour or a statue standing next to a wall, we require semi-cylindrical illuminance (Figure 148).

Figure 148: Example of semi-cylindrical illuminance: 3D view.

The task will be watched from a preferred side, and the fittings can be placed around the suit (see Figure 149), making sure that the shadows cast by each spotlight do not confuse the view.

Figure 149: Example of semi-cylindrical illuminance: plan view.

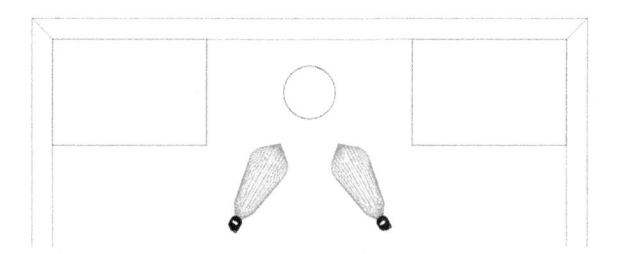

19.3.6 Application: cylindrical illuminance

If the application is similar to the previous one, e.g. a statue, but it is placed in the middle of a room, it requires cylindrical illuminance since people can walk around it. Please note that it is not only the shape of the task that determines the type if illuminance it needs, but also the flow of people using the place (see part 4).

Figure 150: Example of cylindrical illuminance. 3D view, and plan view on the bottom left.

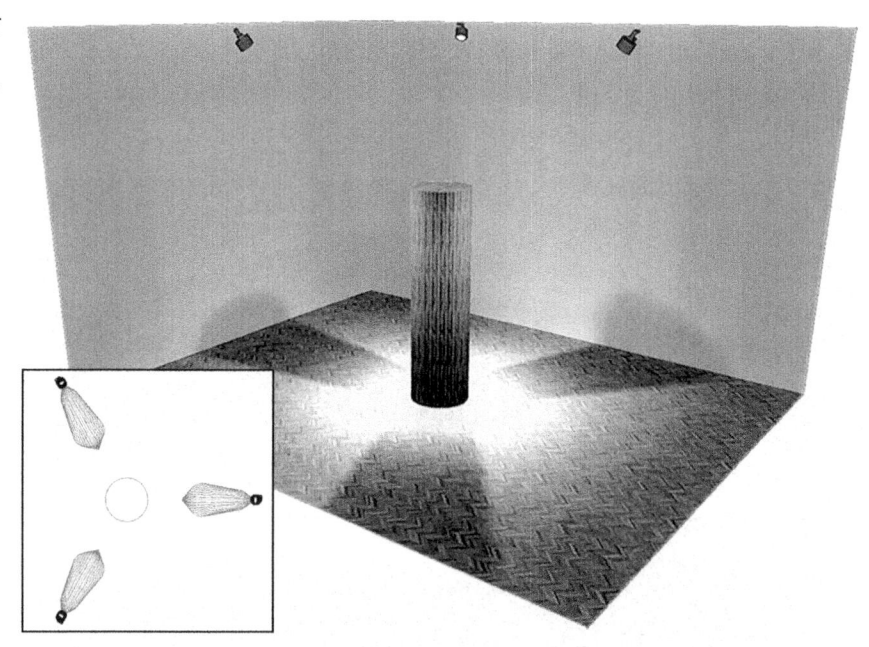

The designer should place at least three luminaires at 120° from each other, all around the task, so that all its sides are lit, and there is no preferred direction of viewing (see Figure 150). The designer must also make sure that none of the lights dazzle the observer.

Both for cylindrical and semi-cylindrical illuminance it may be also necessary to place another trio (at least) of luminaires illuminating the task from the bottom, in order to remove the shadows cast by any protuberance on the task. Imagine that the statue held an open umbrella: illuminating it from the top will mean that a large part of the task will be in shadow, unless one adds a second set of fittings giving light from the bottom up.

In order to avoid an unnatural look, it is important that the bottom set of fittings does not overpower the top one, so that an observer will perceive most of the light coming from the ceiling, not from the floor.

Keep in mind that highly directional lighting causes very dark shadows that can highlight imperfections on the surface of the objects and increase the perceived roughness.

19.3.7 Application: Spherical illuminance

Applications requiring spherical illuminance are usually solved exactly like the ones requiring cylindrical illuminance, with the addition of a source right on top of the task (see Figure 151).

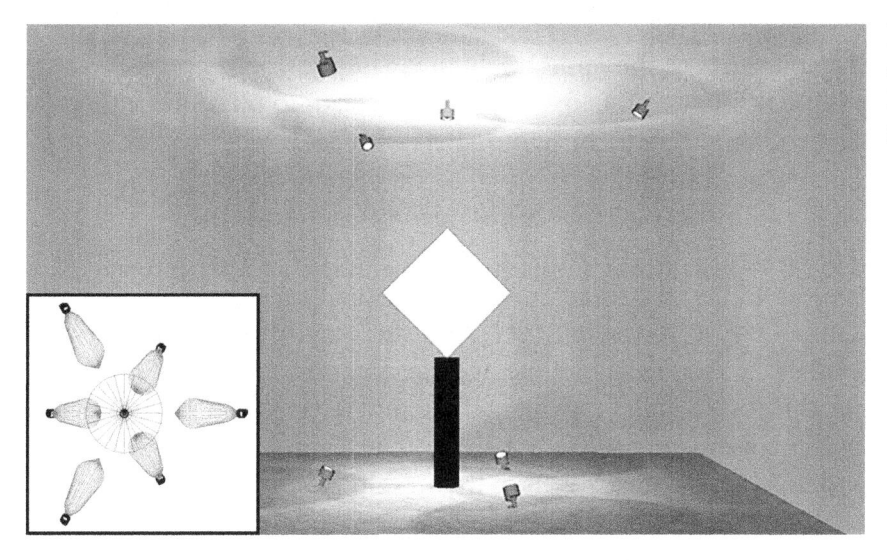

Figure 151: Example of spherical illuminance. 3D view on the left and plan view on the right.

The bottom set of luminaires is almost always necessary here.

19.4 Axially symmetric wide beam

The difference between this fixture and the previous one is that the beam spread is much wider. In the light intensity distribution shown in Figure 152 it is about 80°.

The fitting will create a large round bright spot on the floor or on the work plane, making it useful for the ambient layer (general lighting).

The cone diagram is still useful to give a quick idea of the illuminance and the size of the illuminated area (Figure 152, right side).

Figure 152: Light intensity distribution curve and cone diagram for a wide beam axially symmetric fitting.

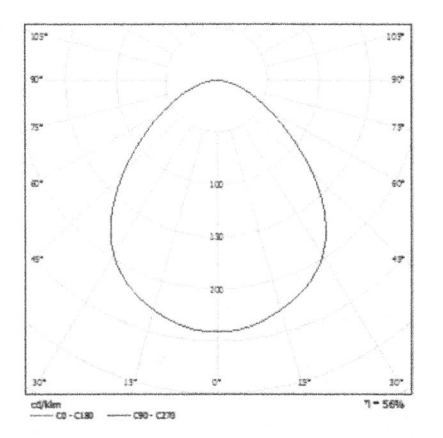

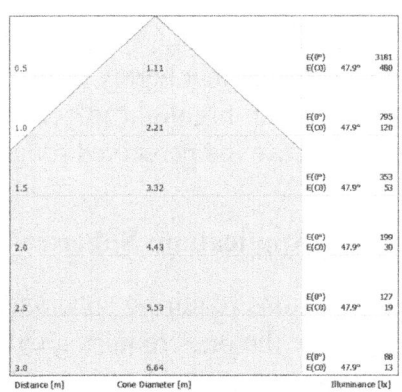

The outskirts of the large bright area (Figure 153 on page 227) are the weak spots of this fixture: it is important to place the fittings so that the bright pools of light match with each other and do not reduce uniformity.

19.4.1 Application: single luminaire in the sample interior

Let us look at the isolux curves for the floor of our sample interior (Figure 154): once again the average illuminance must not confuse us. It is just 40 lux, or about half the previous system, but if we look at the curves themselves we'll see that there is a slow increase in illuminance as we move from the sides of the room to the centre.

Selecting by LID: Axially symmetric

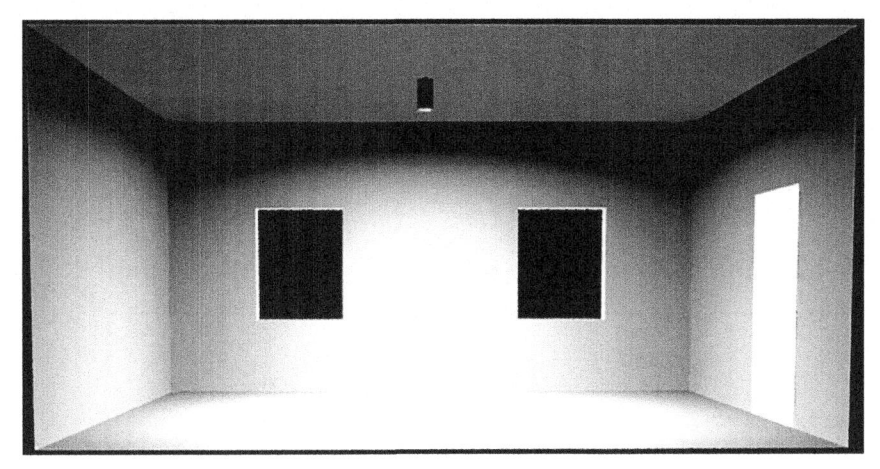

Figure 153: Rendition of the sample interior with axially symmetric wide beam luminaire. Notice the wideness of the pool of light that reaches the walls, and how the floor is not completely illuminated by the fitting.

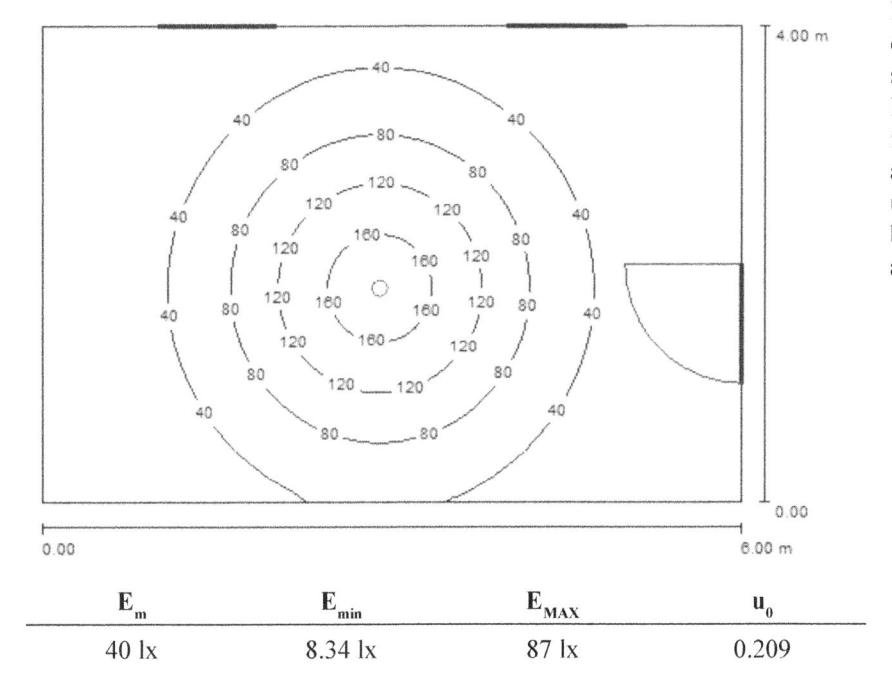

Figure 154: Isolux curves for the axial symmetric wide beam luminaire in the sample interior. Under the diagram, the table of values. Notice the closeness between maximum and average.

E_m	E_{min}	E_{MAX}	u_0
40 lx	8.34 lx	87 lx	0.209

Figure 139 on page 217 showed the behaviour of the narrow beam: the illuminance goes from 200 lx to over 2000 lx in about one meter; with the wide beam, in that same meter, illuminance goes from 20 lx to just 40 lx. This means that there will be much more uniformity than before. The maximum, minimum and U_0 values confirm this.

It is interesting to note that the narrow and wide beam fixtures share the same lamp: a 35 W metal halide.

This type of fixture mounted as a pendant, is very common in residential and hospitality interiors.

19.4.2 Application: sample interior with realistic lighting

Figure 155: Isolux curves for the axial symmetric wide beam luminaire in a real life situation in the sample interior.

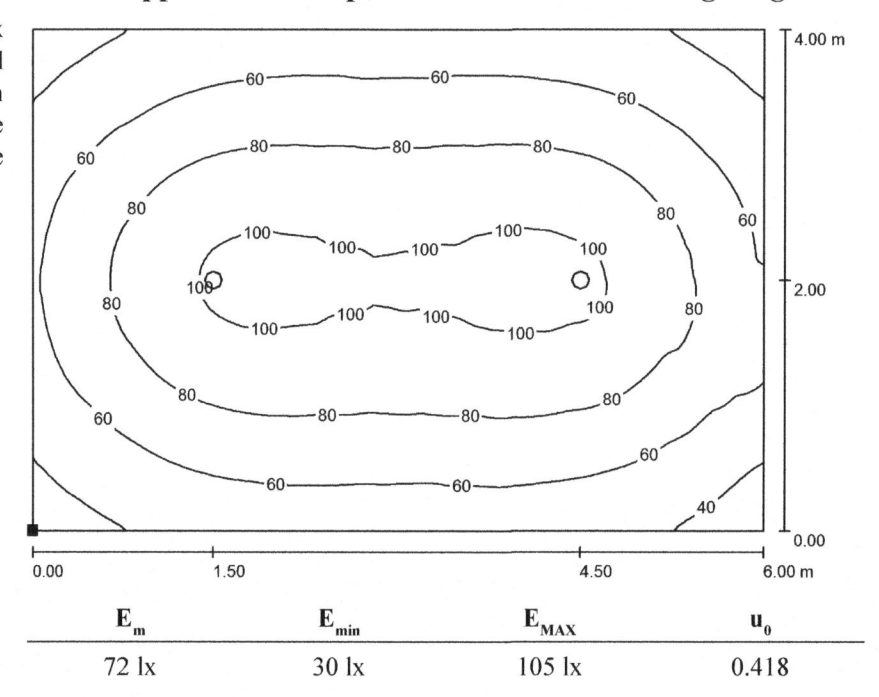

E_m	E_{min}	E_{MAX}	u_0
72 lx	30 lx	105 lx	0.418

If we had to illuminate the sample interior with this type of fitting in real life, we would place two luminaires in a row, so that each could illuminate one half of the room. Each half is almost a square, so the round bright spot lit by the fitting and the indirect light reflected by the walls should cover the room nicely.

Figure 155 shows the isolux diagram for the two-fixture system. The distance between the luminaires is usually half the width of the room (in this case 3 m) and the distance to the wall is half that value (1.5 m).

Depending on the use of the room, we may need to select more powerful fittings (a 70 W metal halide version suggests itself) to provide a higher average illuminance.

19.5 How to dimension an axially symmetric luminaire

This method works for all axially symmetric luminaires, but it is mostly relevant for narrow beam luminaires because it does not consider illuminance uniformity. This is acceptable without significant issues only if the task area is relatively small.

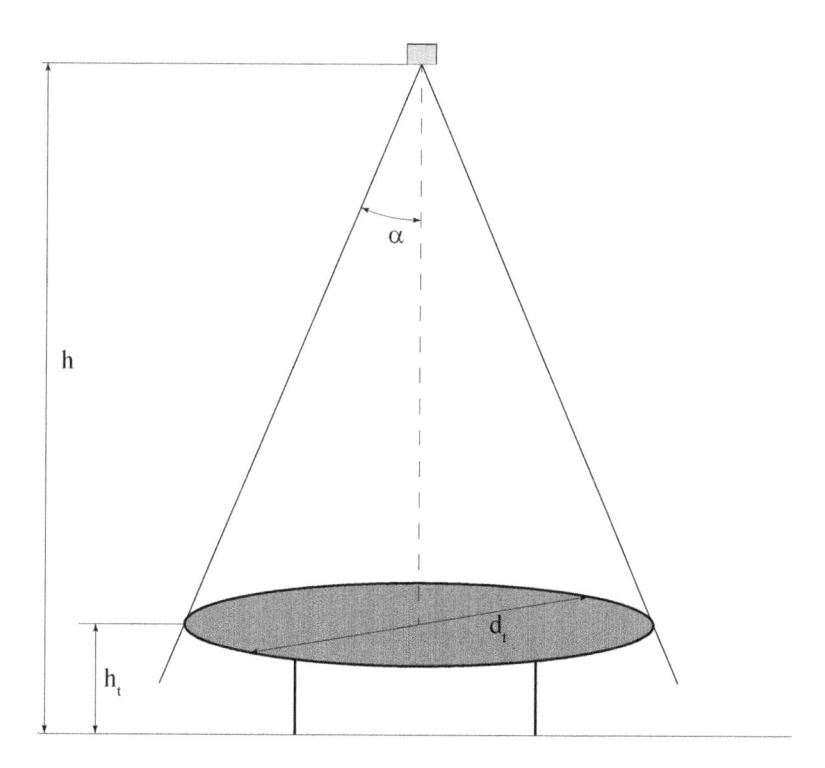

Figure 156: Beam spread used to select the correct spotlight for a task.

There are two main cases we will analyse.

1. When the task is already present and the designer must select the correct luminaire to illuminate it.

2. When the designer has a luminaire already in place and needs to find out whether the task at hand will be illuminated by it.

Let's start with case number one: suppose that we have to select the correct spotlight to illuminate a table in a shop (Figure 142), from a position on the ceiling over the centre of the table. We will measure the height of the ceiling h = 285 cm, the height of the table h_t = 85 cm, and the diameter of the table d_t = 150 cm, then calculate the angle as follows:

$$H = h - h_T = 285 \text{cm} - 85 \text{cm} = 200 \text{cm} \tag{55}$$

$$\alpha = \arctan \frac{\frac{d_T}{2}}{H} = \arctan 0.375 = 20° \tag{56}$$

$$Beam = 2\alpha = 40° \tag{57}$$

Now we know that the correct spotlight should have a beam spread of 40°.

The second step is to find out the intensity it needs to emit: from the activity performed in the interior we get the illuminance required; with this information we can find the correct cone diagram for the luminaire, or calculate its luminous intensity and compare it to its light intensity distribution curve.

Alternatively (case number two) suppose that we are trying to find out how large is the task that can be lit by a spotlight.

Still referring to Figure 156, this time we know h, h_T and α and we need to calculate d_T. The formula is:

$$d_T = 2(h - h_T) \tan \alpha \tag{58}$$

Done that, we can either calculate or measure (recommended in case of old systems) the illuminance the luminaire produces, and, comparing this with the level required by the regulations, find out whether the luminaire works for the task at hand.

19.6 Self evaluation

Answer the following questions to test your understanding of the material:

1. Describe axial symmetric narrow beam luminaires: their light intensity distribution, typical light sources, behaviour, ideal tasks, examples of luminaires, examples of systems and the type of illuminance these generate.

2. Describe a lighting system for vertical illuminance, one for horizontal, one for spherical, one for cylindrical, one for semi-cylindrical. Give details on the types of light intensity distributions, the fittings and their positioning.

3. Describe axial symmetric wide beam luminaires: their light intensity distribution, typical light sources, behaviour, ideal tasks, examples of luminaires, examples of systems and the type of illuminance these generate. Apply this luminaire to a real sample interior and discuss the results.

4. Show how to dimension an axially symmetrical narrow beam luminaire for a circular task with the mounting point right over the centre of it.

5. Show how large a task will a luminaire light when the former is placed right under it.

Figure 157: Sample interior to show the results of different types of fittings.

Figure 158: Sample room - plan view.

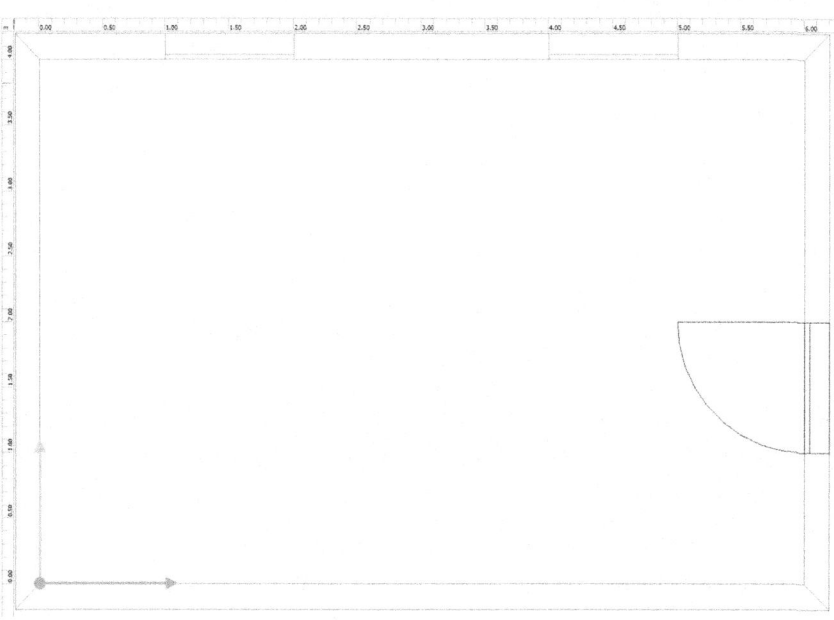

Selecting by LID: symmetric about two planes

20

This kind of fixture emits light very differently crosswise and lengthwise, usually because of a long and narrow lamp (traditional fluorescent for example).

To go on with the kitchen metaphor, the photometric solid is a bit like a loaf of bread: the slices are very different if one cuts lengthwise or crosswise.

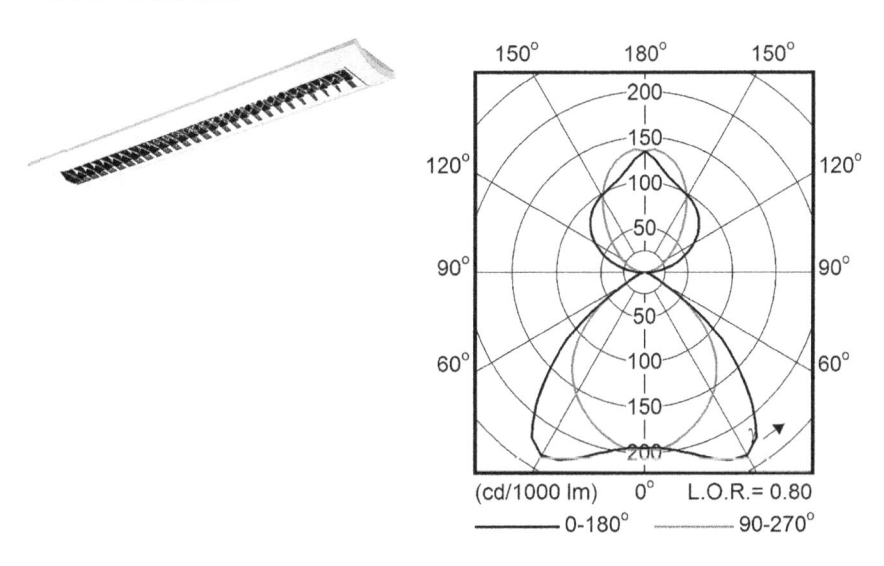

Figure 159: Direct-indirect fitting. From left to right: a picture of the luminaire and its light intensity distribution: it is symmetric about two planes.

The advantage of these fixtures is that the high luminance spot on the floor looks more like an ellipse, bordering on the rectangular, so it is better suited to rectangular rooms.

20.1 Symmetric about two planes: narrow beam

Usually the fixtures that belong to this category are outdoor projectors that employ a powerful dual ended discharge lamp, compact or otherwise, and have no real use in interiors.

They are of no real interest to the book, so I will say no more about them.

20.2 Symmetric about two planes: wide beam

This is a very common fixture, it comes in a variety of mountings (pendant, ceiling mounted, recessed, track mounted, even floor lamps) and it is mostly used to provide general lighting.

Figure 160: Light intensity distribution and photometric solid of a light fitting symmetric about two planes.

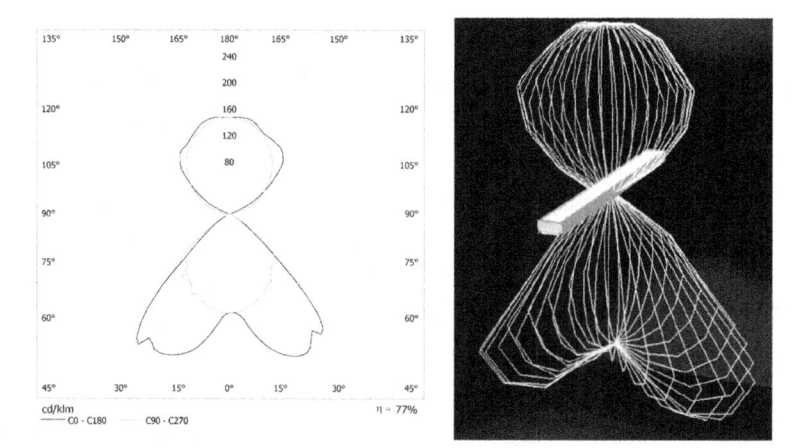

20.2.1 Application: single fitting in the sample interior

The direct photometric curve shows the asymmetry of emission of the fixture: crosswise (in red on the left part of Figure 160) the beam spread is much wider than lengthwise (in blue on the left part of Figure 160).

In the particular luminaire I selected for this example, the indirect light intensity distribution is almost perfectly axially symmetric. This means that we can compare the effect of these two types of light intensity distributions by just looking at the ceiling and the floor in the sample interior.

Selecting by LID: symmetric about two planes

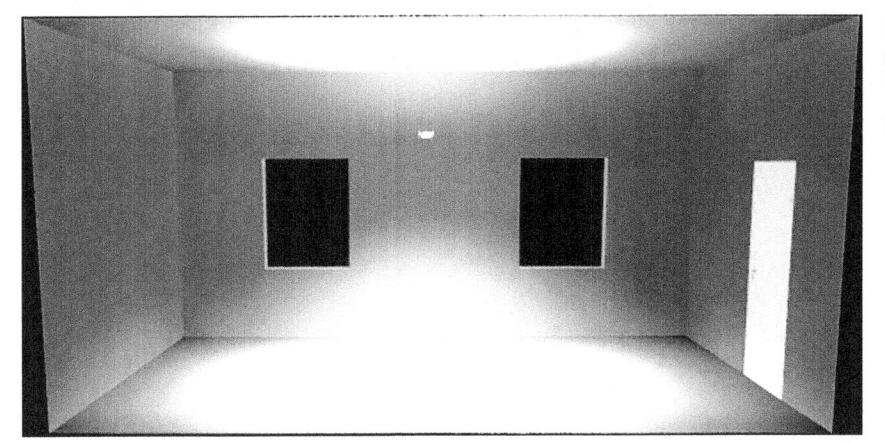

Figure 161: Render of the sample interior with luminaire symmetric around two planes.

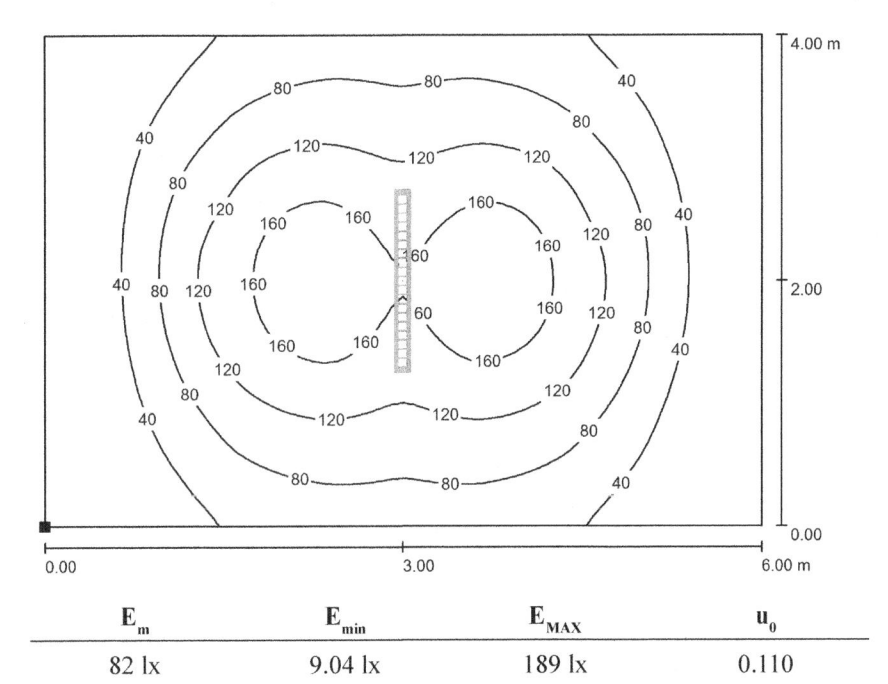

Figure 162: Isolux on the floor for the luminaire symmetric about two planes in the sample interior.
Under the diagram, the table of values. .

E_m	E_{min}	E_{MAX}	u_0
82 lx	9.04 lx	189 lx	0.110

The shape (Figure 161) of the bright areas on floor and ceiling follow the shape of the light intensity distributions: there is a large elliptical pool of light on the former and a round one on the latter. This shows shows the advantage of the symmetric about two planes over the axially symmetric result on the ceiling ("Figure 163: Isolux on the ceiling for the luminaire symmetric about two planes." on page

236): it is better at following the shape of a rectangular room. This effect is especially evident looking at the 80 lx isoline.

Figure 163: Isolux on the ceiling for the luminaire symmetric about two planes.

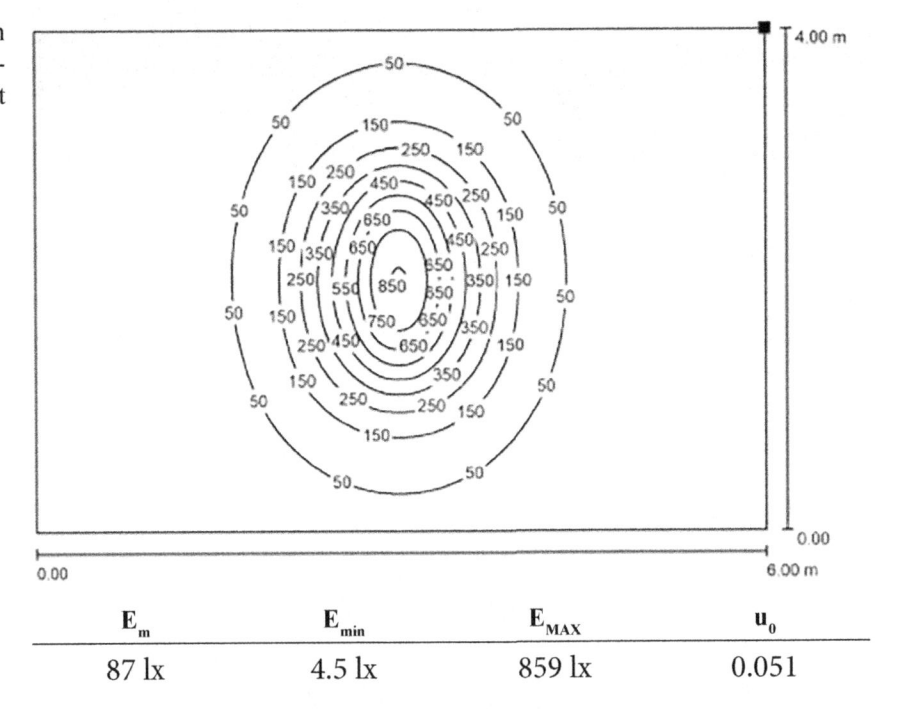

E_m	E_{min}	E_{MAX}	u_0
87 lx	4.5 lx	859 lx	0.051

Studying carefully the render, we notice two further issues.

1. The lengthwise direct light distribution is slightly too wide for the room, because while the crosswise walls are dark, the lengthwise ones are partially lit.

2. There are actually two bright spots on the floor, on the sides of the fitting, rather than just a wide one. The reason for this is that the shape of the direct light intensity distribution of this luminaire is not exactly ideal: the two half-curves are too close together. This means that some intensities are so high that they damage illuminance uniformity in the areas described by the 160 lx isoline (Figure 162).

The uniformity is rather bad ($U_0 = 0.110$). The reason for this is that a single luminaire is not enough to cover the whole of the room. The minimum illuminance is very low ($E_{min} = 9.04$ lx), but even so the

Selecting by LID: symmetric about two planes   237

average is almost acceptable, especially if one considers the effect on it that the area beyond the 40 lx curve will have.

This is not a surprise: the idea was to place a fitting in an interior slightly too large for it, so that we could truly appreciate its performance and its limits.

20.2.2 Application: Sample interior with realistic lighting

In a real life situation we would use two fittings in the centre, oriented parallel to the longer walls, so that each can take care of half the room (Figure 164).

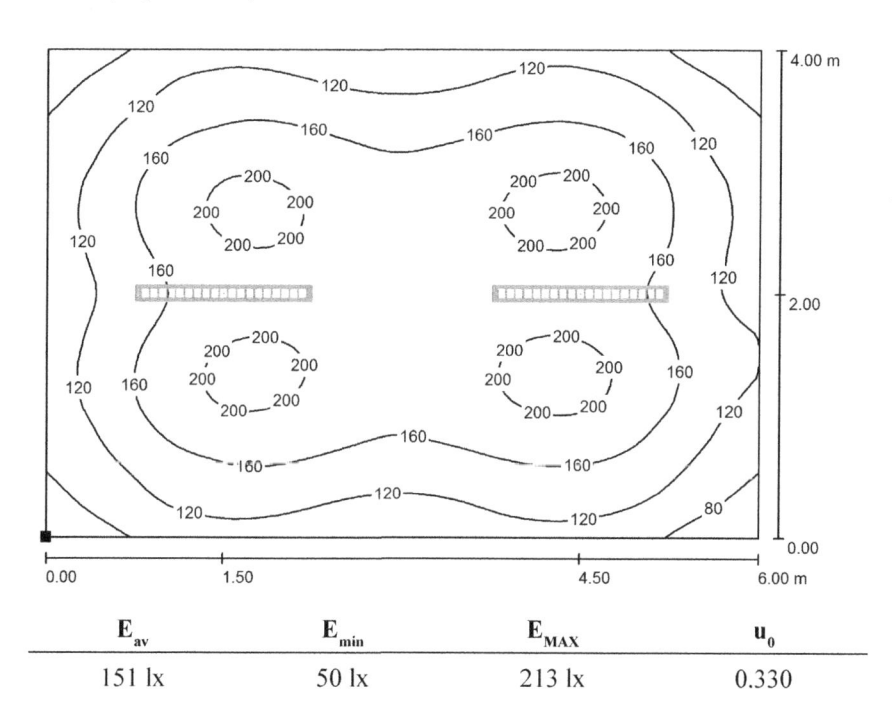

Figure 164: I s o l u x curves for the luminaire symmetric about two planes, in a real life situation in the sample interior.

E_{av}	E_{min}	E_{MAX}	u_0
151 lx	50 lx	213 lx	0.330

This system provides 155 lx on average and its uniformity rises to 0.330, which is acceptable, because the reflectance of the walls is unrealistically low; with a more realistic value for it, we would notice a significant increase in minimum illuminance. The distance between maximum and average illuminance shows that the fitting is not a fantastic performer, because of the imperfection in its light intensity distribution.

20.2.3 How to select the right orientation

In order to emphasise the importance of the orientation of these fittings, let's select a different interior and luminaire: a six metre by two-metre corridor.

Figure 165: I s o l u x curves for the luminaire symmetric about two planes in the sample hallway. Longitudinal orientation.

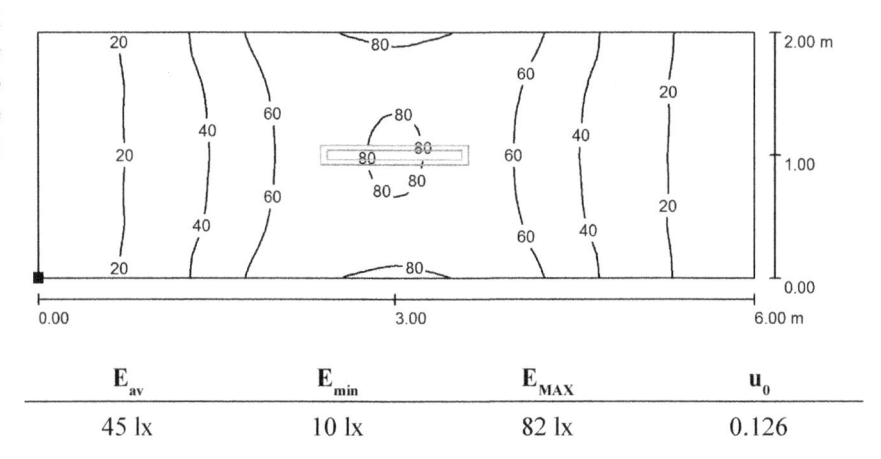

E_{av}	E_{min}	E_{MAX}	u_0
45 lx	10 lx	82 lx	0.126

Let's study the two possible placements.

Figure 166: I s o l u x curves for the luminaire symmetric about two planes in the sample hallway. Transversal orientation.

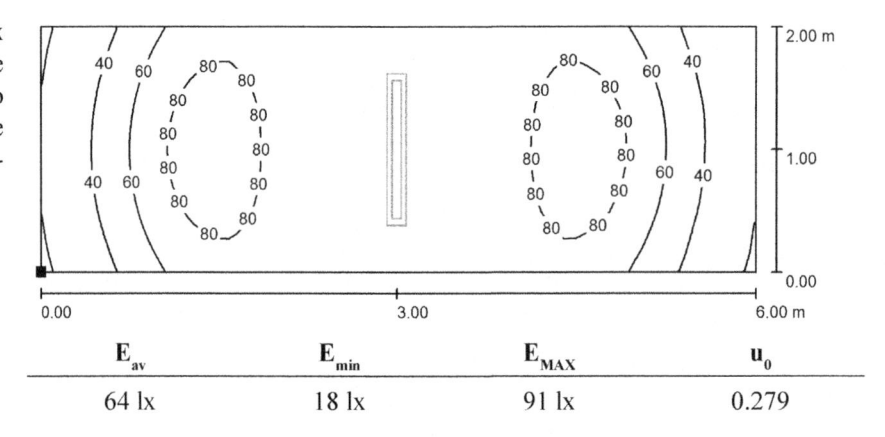

E_{av}	E_{min}	E_{MAX}	u_0
64 lx	18 lx	91 lx	0.279

The longitudinal placement (Figure 165), with respect to the hallway, is possibly the most intuitive: the long side of the fitting is aligned to the long side of the room. However, even a quick look at

Selecting by LID: symmetric about two planes

the isolux shows the issue: the maximum values of illuminance are in the centres of the longitudinal walls, instead of being on the floor, and the uniformity in the direction of the hallway is rather bad.

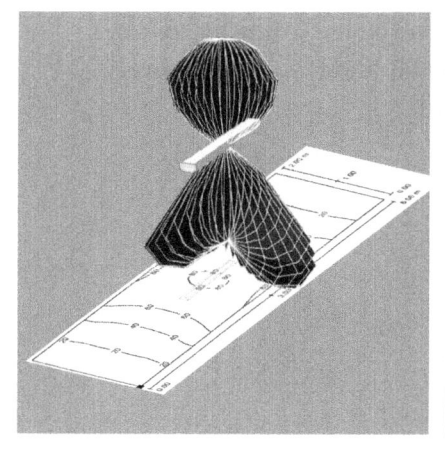

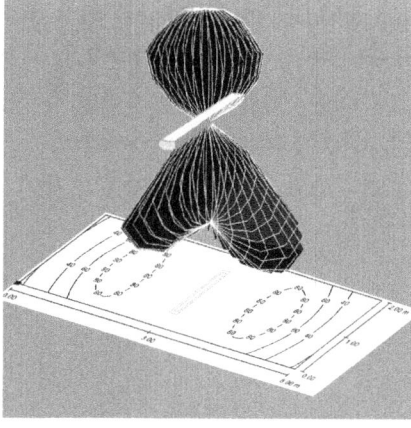

Figure 167: Incorrect (left) and correct (right) positioning of the luminaire according to the light intensity distribution.

The crosswise placement of the fitting (Figure 166) has its maximum illuminance along the way, and it appears to be the correct one.

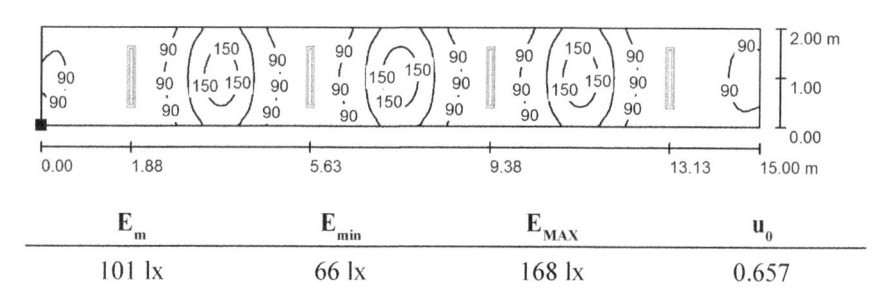

Figure 168: Isolux curves for the luminaire symmetric about two planes, in a real life situation in the sample hallway.

E_m	E_{min}	E_{MAX}	u_0
101 lx	66 lx	168 lx	0.657

Let's compare results and see whether the numbers confirm our hypothesis: The average illuminance rises by 42%, the minimum by 80%, the maximum by 11% and the uniformity by 23%. This improvement in performance is impressive, considering that it comes from a simple 90° rotation of the fitting.

This is the practical demonstration of the importance of studying the light intensity distribution curve when building a lighting system (59).

> The **right use** of a luminaire for the task
> derives from the study of the **light inten-** (59)
> **sity distribution curve.**

In a real life installation, we would place the fittings at about 3.75 m of distance from one another, to obtain the result in Figure 168.

20.2.3.1 Luminance control

Figure 169: Luminance diagram calculated by DIALux.

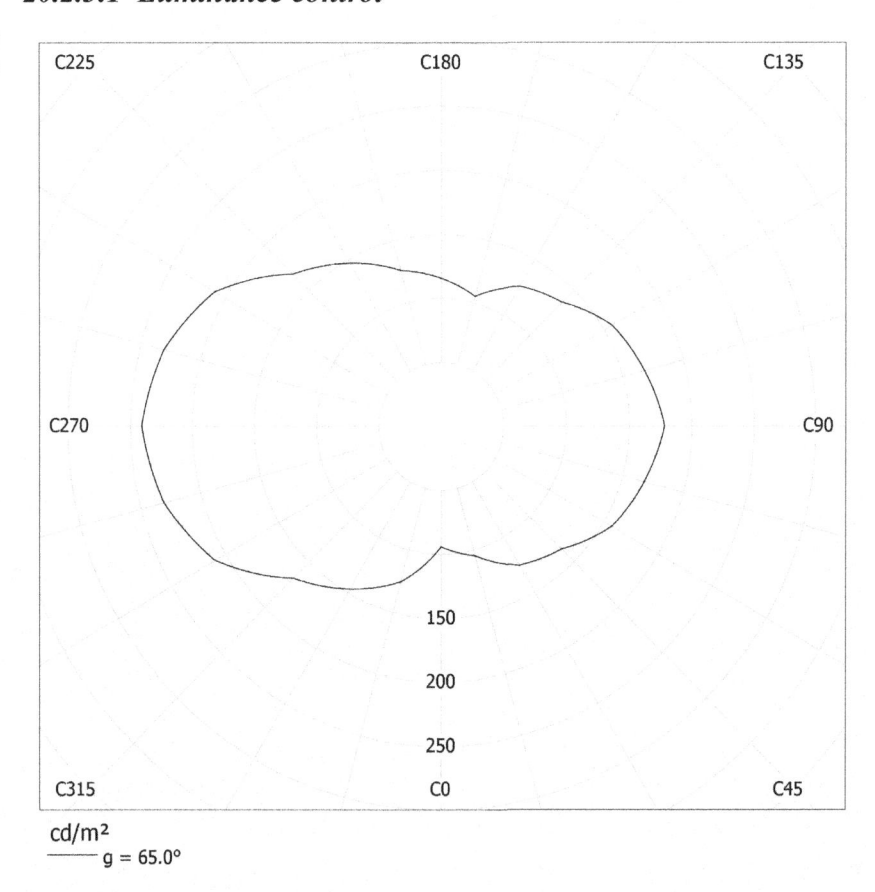

The fixtures I used in the examples of this section are fluorescent suspended luminaires. When they are employed in offices or places where there are computer monitors, they must comply with luminance regulations[1]. The manufacturer will declare compliance, or the designer will use the luminance diagram (Figure 169) and the luminance table from DIALux to make sure: in the example shown, the maximum luminance is under 250 cd/m^2, this means that it can be used in any interior with DSE ("9.4.2 Luminance control for workstations with Display Screen Equipment" on page 109).

20.3 Self evaluation

Answer the following questions to test your understanding of the material:

1. Describe wide beam symmetric about two planes luminaires: their light intensity distribution, typical light sources, behaviour, ideal tasks, examples of luminaires, examples of systems and the type of illuminance these generate. Apply this luminaire to a real sample interior and discuss the results.

2. Describe how to select the correct orientation of a symmetric about two planes luminaire in an interior.

---

1. In Europe, this means complying to the EN12464-1:2011 standard

Figure 170: Sample interior to show the results of different types of fittings.

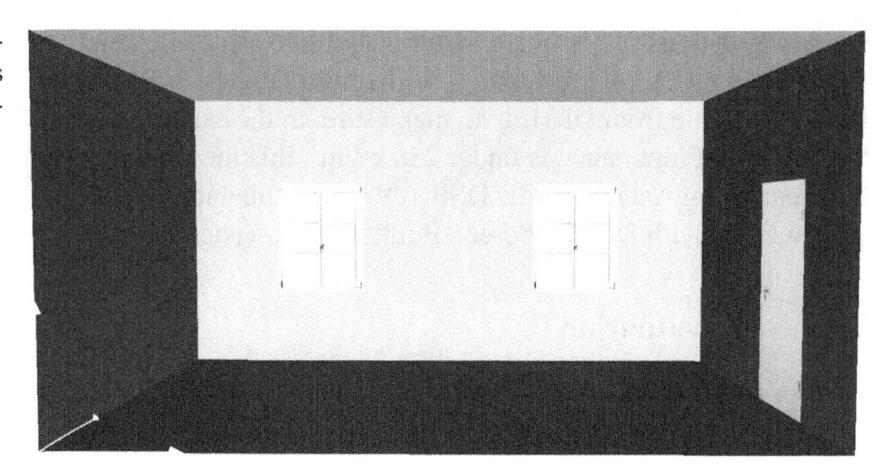

Figure 171: Sample room - plan view.

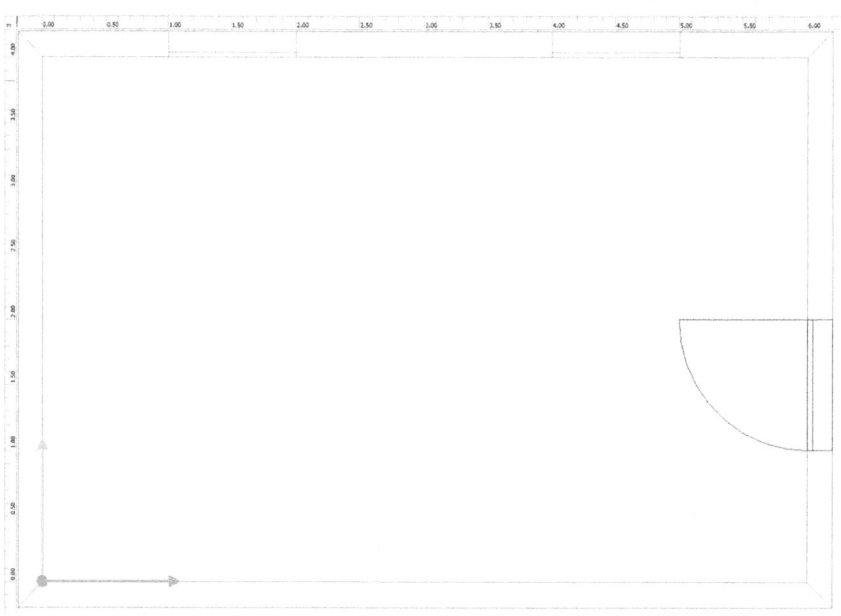

Selecting by LID: asymmetric

21

The name of this type of luminaire comes from the asymmetric (usually crosswise) light intensity distribution curve they have.

Looking at Figure 173, the dark grey curve is crosswise and asymmetric, while the light grey lengthwise one is symmetric.

The light intensity distribution solid is shown in Figure 172, and it is conceptually similar to a banana: slicing it crosswise for a fruit salad yields a much different result than slicing it lengthwise for a banana split. There is a plane of symmetry: cutting the banana lengthwise will create two symmetric halves.

The advantage of these fixtures is that they can be placed off to the side of a task, so as to be out of sight, and still illuminate it. If the task is a frescoed wall or a painting hanging from a wall, the ideal position for a symmetric fixture is in the centre of the painting, obstructing the view. On the contrary, an asymmetric can be placed above the painting on a small strut and illuminate the whole of the task without obstruction.

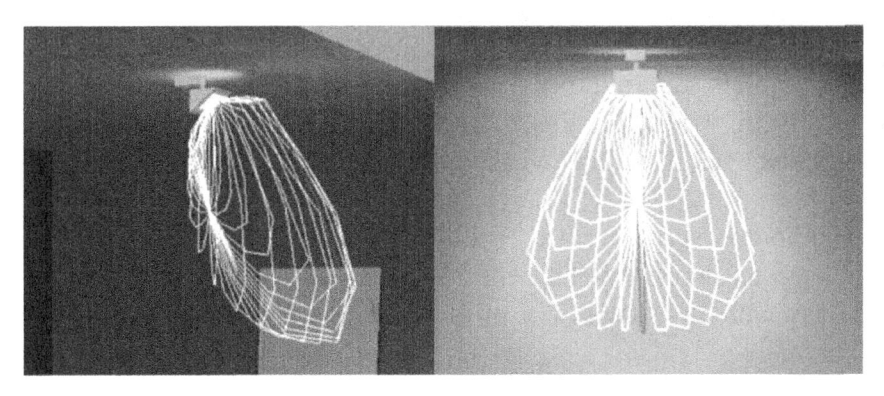

Figure 172: Light intensity distribution solid for an asymmetric light fitting.

The ideal asymmetric curve is of course half of the ideal photometric curve, oriented differently (see following paragraphs).

Unfortunately with this type of fitting it is even harder to find a fixture that comes close to the ideal curve.

21.1 Traditional asymmetric

Figure 163 shows a sketch of the behaviour of this type of fitting and an example of light intensity distribution curve. The luminaire is placed on the side of the task, and the angular size of the usable part of the photometric curves defines the task that can be illuminated.

Looking at the crosswise curve (blue in Figure 173) we can deduce the behaviour of the fitting.

Figure 173: Sketch and photometric curve of a traditional asymmetric light fitting.

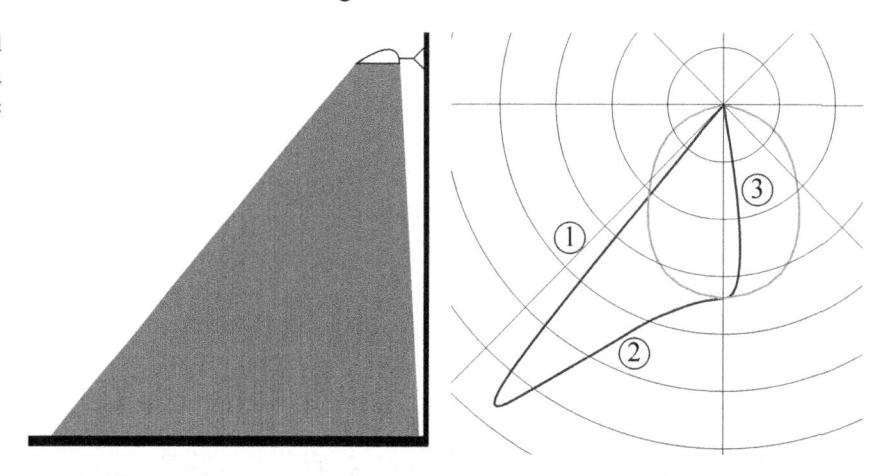

The top left part (number 1) is flux dispersed away from the task. Ideally we would like this curve to be as straight as possible, to be efficient and not damage visual comfort.

The bottom side (number 2) follows the ideal shape from gamma = 0 up to about 40°. This means that it can uniformly illuminate a task with an angular size of 40° from the mounting point of the luminaire.

The right side (number 3) shows how much light goes beyond the axis of the luminaire. The rounder the line, the higher the flux sent in these directions. Because these fittings are usually placed like in the

Selecting by LID: asymmetric

sketch in Figure 173, this usually means wasted flux, so we should prefer straighter curves.

The lengthwise curve (red in Figure 173) determines the width of the task.

These fittings are very commonly used to illuminate everything from sport arenas to streets to large outdoor areas. It is not so frequent to see them indoors: even when they might be a better solution, since there is a tendency by inexperienced lighting designers to select a symmetric luminaire.

Figure 174: Asymmetric luminaire lighting a ceiling.

Asymmetric fittings are often used to illuminate ceilings (see Figure 174), and provide general indirect illumination or diffuse lighting to a room. To minimise the risk of glare on the wall, they are mounted as high as possible. The problem with this is that the beam spread needed to cover the same ceiling area increases with the mounting height. If the fitting is unable to provide it, there will be another dark, low luminance spot, this time on the ceiling.

21.1.1 Application: single fitting to sample interior

The fitting is mounted on top of the wall, aimed downwards.

Figure 175: Render of the sample interior with asymmetric luminaire.

Figure 176: False colour display indicating luminance for the sample interior with asymmetric luminaire.

The first thing to note in the render in Figure 175 is that there is a relevant quantity of flux ending on the wall itself. This has two drawbacks: the actual efficacy of the system will decrease because there is luminous flux not going onto the task and, with a high enough reflectance, the wall might dazzle the users of the room.

Selecting by LID: asymmetric

It is necessary to make sure that the luminance contrast is not large enough to cause glare, and this can be done, although in a very approximate way, via DIALux's luminance false colour display (Figure 176).

We will see in paragraph 21.7.3.1 on page 466 that the luminance output in DIALux works only if all the surfaces are perfectly diffuse. If this is not so, the designer might use a false colour illuminance display and manually evaluate the reflectances of the different surfaces.

Figure 177 shows the isolux on the floor.

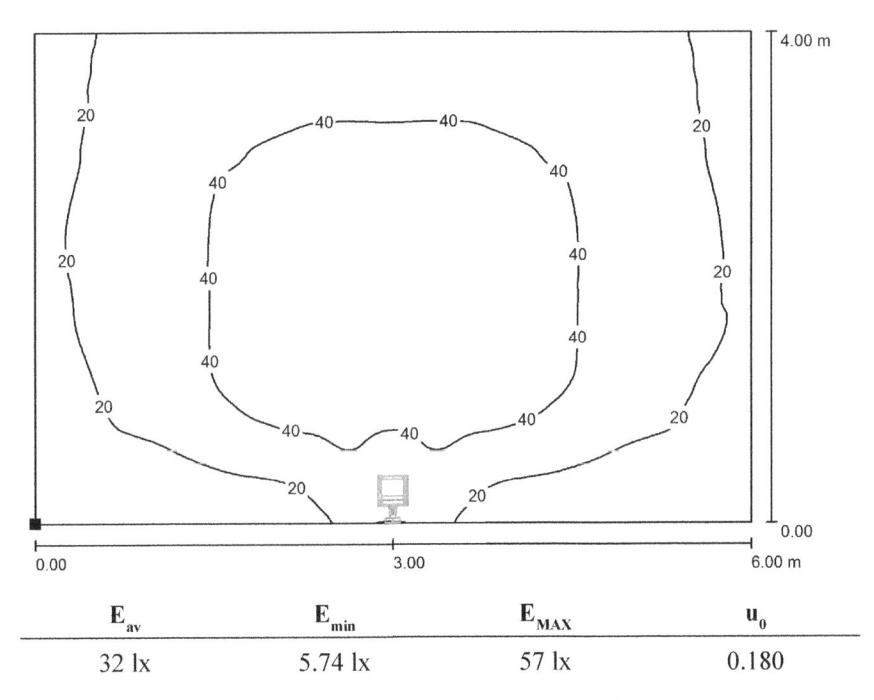

Figure 177: Isolux curves for the asymmetric luminaire on the floor of the sample interior.

E_{av}	E_{min}	E_{MAX}	u_0
32 lx	5.74 lx	57 lx	0.180

The single fitting is not able to cover the whole floor. The interesting thing is that the average illuminance (32 lx) is close to the maximum (57 lx): this means that the luminaire is good at providing uniformity… where it can reach. Sure enough the uniformity factor is very low ($U_0 = 0.180$) because, again, the room is too large for a single fixture.

21.1.2 Application: Sample interior with realistic lighting

In a real life situation (see Figure 178) the idea is to use more luminaires placed the exact same way, not to waste the uniformity they provide. Figure 167 shows the isolux: the average went up to 113 lx, with a good uniformity ($U_0 = 0.465$). As usual, a room with a higher reflectance will have even better results.

Figure 178: Isolux curves for the asymmetric luminaire in the sample interior. Real life example.

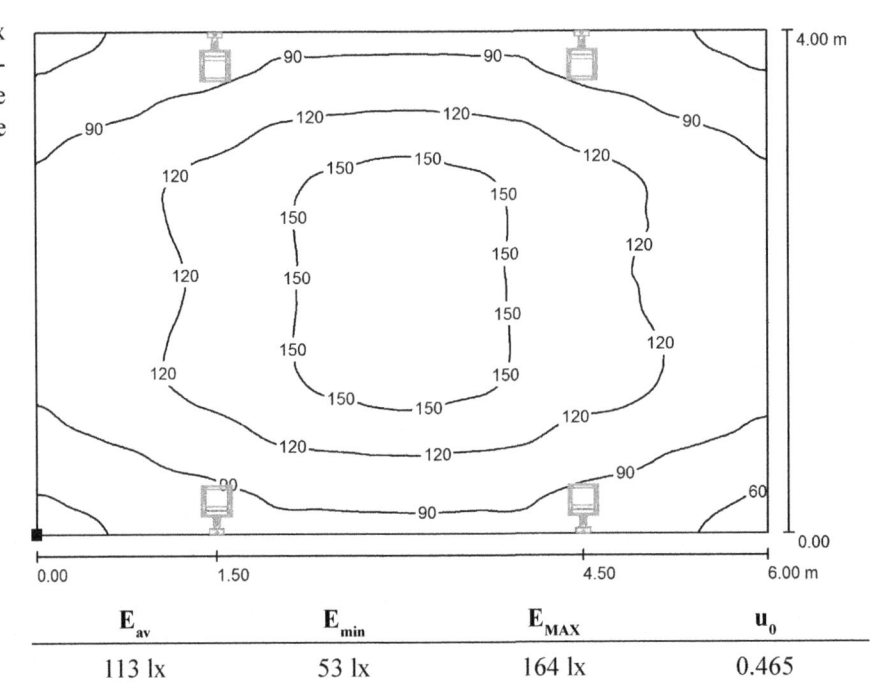

E_{av}	E_{min}	E_{MAX}	u_0
113 lx	53 lx	164 lx	0.465

With four fittings it is even more important to verify that the luminance contrasts on the walls aren't damaging the room's visual comfort.

21.2 Wall washers

Figure 179 shows a sketch of the wall-washer fixture and its light intensity distribution. It shows the difference between a traditional asymmetric and a wall-washer: the former has its opening parallel to the task, while the latter has the opening perpendicular to the task. This means that the wall-washer can be placed flat on a ceiling and illuminate a wall, thus almost disappearing from view. Some are even recessed into false ceilings.

It should be noted that the ideal part of the light intensity distribution curve in the wall-washer is a 45° mirrored version of the traditional asymmetric because the former has a vertical task while the latter has an horizontal one.

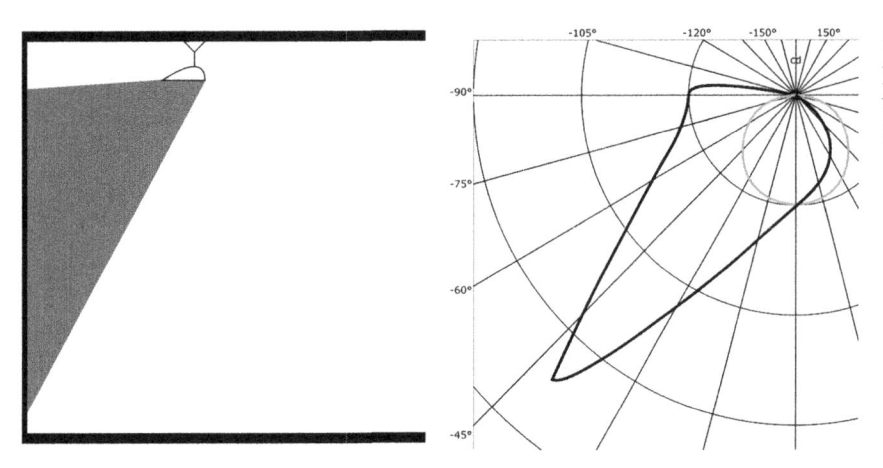

Figure 179: Sketch and photometric curve of a traditional asymmetric light fitting.

The main problem with this type of fitting is illuminating the top of the wall, near the ceiling.

Looking at the crosswise light intensity distribution on the right side of Figure 179 we can deduce some facts about the wall-washer.

1. The dark grey curve describes the asymmetric behaviour of the luminaire: the left side (number 1 in the picture) is the ideal distribution, up to about 40° in our example. It must be as close to the ideal photometric curve as possible.

2. The bottom side (number 2 in the picture) shows the intensities being dispersed on the floor. Ideally one would like the curve to be as straight as possible, back towards the origin, to reduce wasted luminous flux.

3. Finally the right side of the curve (number 3 in the picture) shows what happens on the opposite side of the task. This can damage visual comfort, as it is intensity sent in the direction of

the user of the task. Please note that this part of the curve may not be there at all, see for example Figure 183.

4. The light grey curve in Figure 179 determines the width of the beam. Again this curve may disappear if the crosswise curve is wholly contained into one quadrant.

We can identify two types of wall-washers according to the width of the lengthwise beam: a narrow and a wide version.

21.2.1 Application: narrow beam wall-washer and sample interior

Figure 180 shows the photometric solid of the luminaire (on the left) and a photometric curve on the right.

There is actually little asymmetry in the fitting itself, some may even behave like spotlights tilted by a few degrees in the C0-C180 plane.

Obviously the task these fittings illuminate is narrow and tall.

Figure 180: Sketch and photometric curve of a narrow beam wall-washer light fitting.

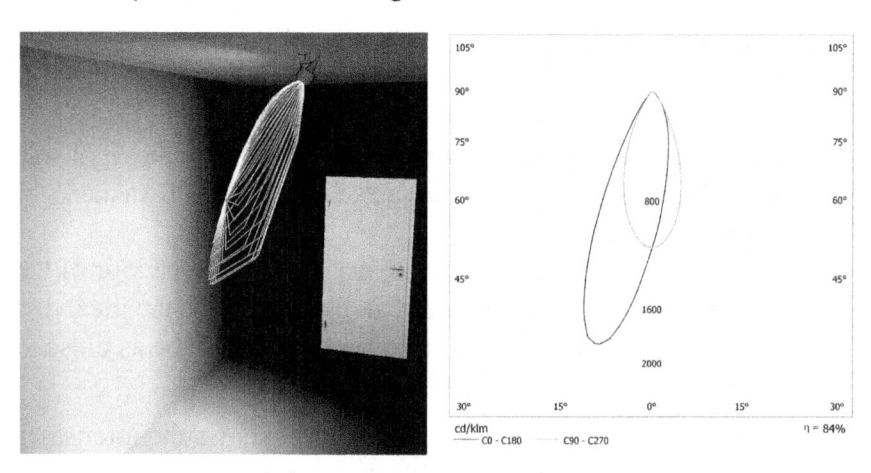

Figure 181 shows a visual render of the bright spot this light fitting generates in the sample interior, and Figure 182 provides the isolux curves and the value of illuminance on the wall.

When placing these luminaires it is important to avoid excessive luminance contrasts on the walls caused by the narrowness of the

Selecting by LID: asymmetric

beam. To reduce this effect, the designer needs to select a less powerful fitting to reduce the amount of luminous flux ending on the wall.

Figure 181: Render of the sample interior with a narrow beam wall-washer luminaire. Here, the task is the wall.

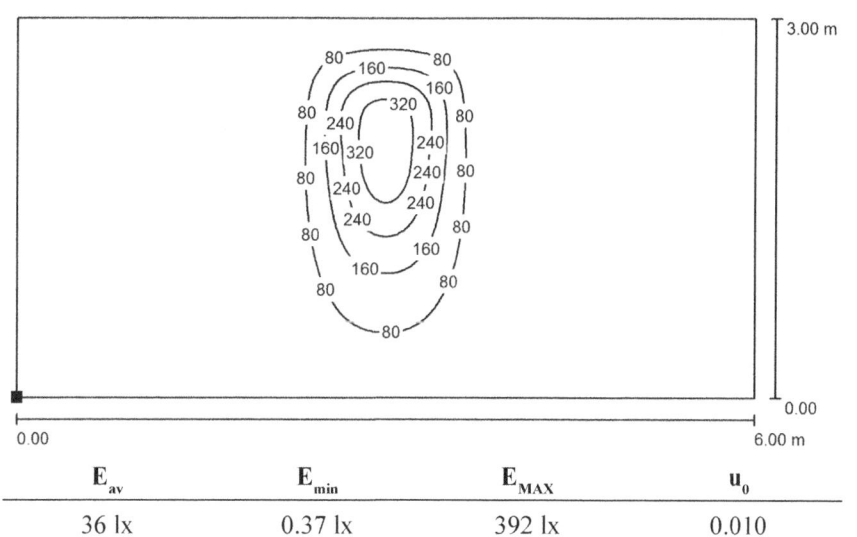

Figure 182: Isolux curves for the narrow beam wall-washer luminaire in the sample interior.

E_{av}	E_{min}	E_{MAX}	u_0
36 lx	0.37 lx	392 lx	0.010

Another issue to avoid is placing or orienting the luminaire so that there is a spot of high illuminance on the floor in front of the task. This would highlight an irrelevant area of the interior, reduce visual comfort, waste luminous flux and ruin illuminance uniformity on the floor. To solve this problem, the designer needs to move the fitting closer to the task.

21.2.2 Application: wide beam wall-washer and the sample interior

Figure 183 shows an example of a wide beam wall-washer with its photometric curve, and Figure 172 shows it installed in our sample interior. Again its task is the north wall.

With this type of wall-washer some of the light reaches the floor and the walls on the side, missing the task. This is a much less damaging issue than it was with the narrow beam, as illuminance here increases slowly and smoothly, so visual comfort will be much less affected. The high luminance spot is now on the ceiling: much less visible than the ones the asymmetric caused.

Obviously these fitting illuminate a wide task well.

Please note that illuminating a wide task with a narrow beam wall-washer will be as awful as providing general lighting with a narrow beam spotlight, because the result will either be a series of alternating bright and dark vertical stripes on the walls, with terrible visual comfort, or a grossly over-lit wall that again will destroy any feeling of well being in the room.

Figure 183: Wall-washer fitting and its light intensity distribution.

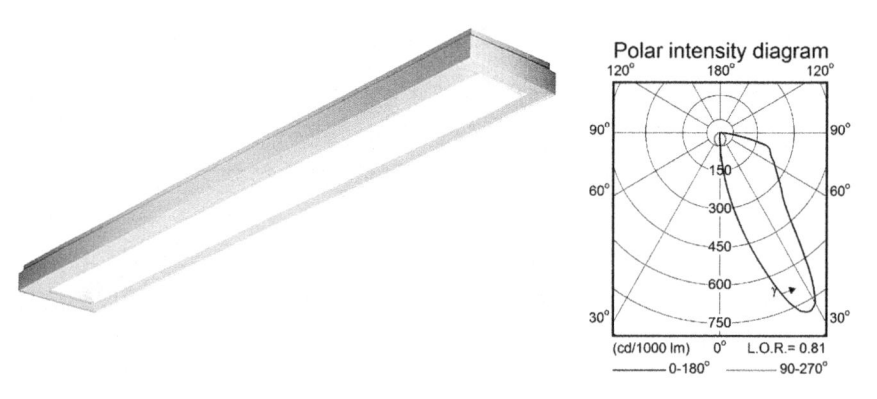

Observe that in Figure 183 there is just one light intensity distribution curve. The reason for this is that the luminaire has no emission on the lengthwise vertical plane. If we tilted the plane of measurement by 30°, we would see the width of the longitudinal emission of the luminaire.

Figure 184 and Figure 185 show that a single fitting illuminates part of the wall in a uniform way: the illuminance goes from 20 lx to 80 lx in two metres.

Figure 184: Render of the sample interior with a wall-washer luminaire. Here, the task is the wall.

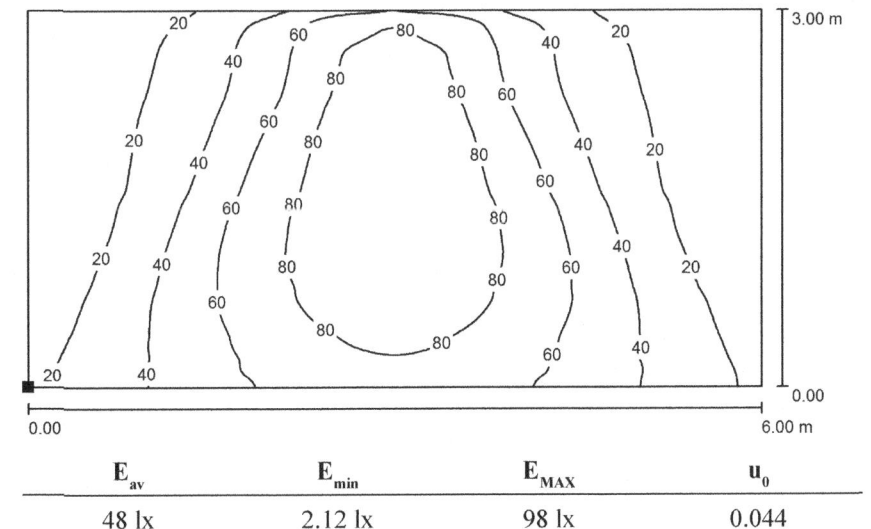

Figure 185: Isolux curves for the wall-washer luminaire in the sample interior.

E_{av}	E_{min}	E_{MAX}	u_0
48 lx	2.12 lx	98 lx	0.044

Asymmetrics can glare if they are oriented incorrectly (see Figure 186): it is important to make sure that the right side of the fitting is facing the task both while designing and while actually installing the fixtures, especially for recessed ones.

Figure 186: Example of wrong (left) and (right) orientation of a wall-washer luminaire when the task is the north wall.

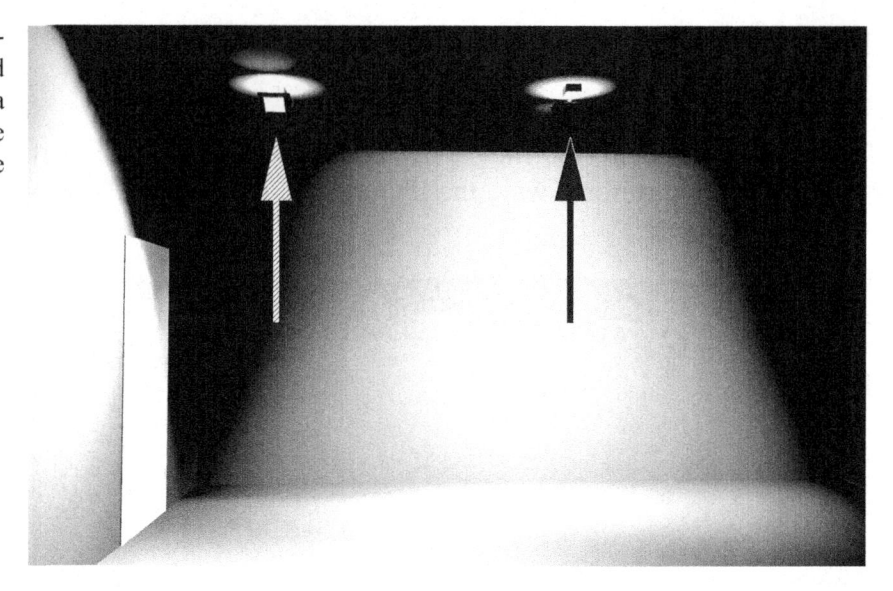

21.2.3 Application: Sample interior with realistic lighting

Let's apply this luminaire to a more realistic situation: we want to illuminate the whole wall with as few luminaires as possible.

Figure 187: Isolux curves for the wall-washer luminaire in the sample interior. Real life example.

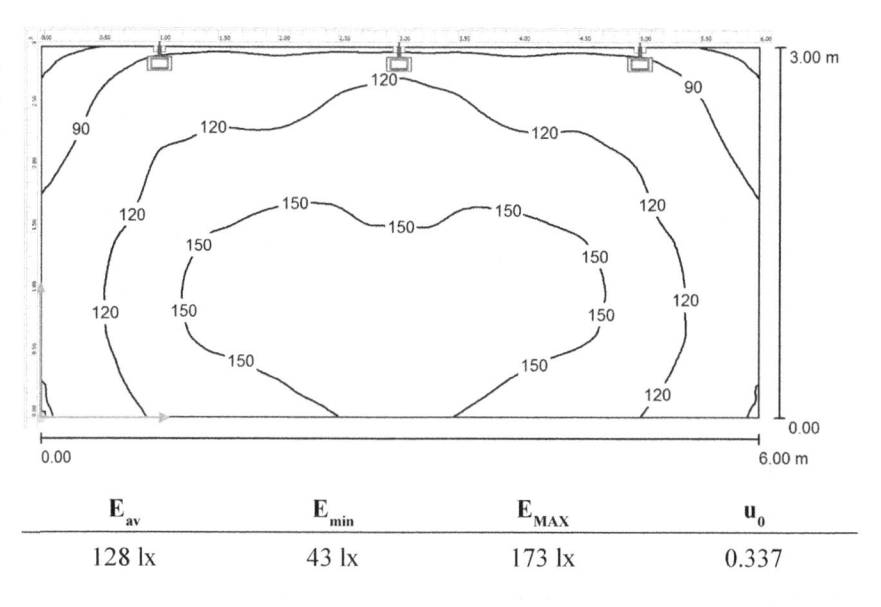

E_{av}	E_{min}	E_{MAX}	u_0
128 lx	43 lx	173 lx	0.337

We need to use three wall-washers (Figure 187), evenly spaced, so that the distance between them is twice as much as the distance between the east (or west) wall and the first (or last) luminaire. The wall is now completely lit, with an acceptable, if low, uniformity of 0.337. The value is kept low by the very low reflectance of the walls.

21.3 Fully asymmetric luminaire

These fixtures are often purely decorative (see Figure 188). The light intensity distribution has no symmetry, and they are usually quite uninspiring performance wise, even though some of them can be forced into one of the previous categories.

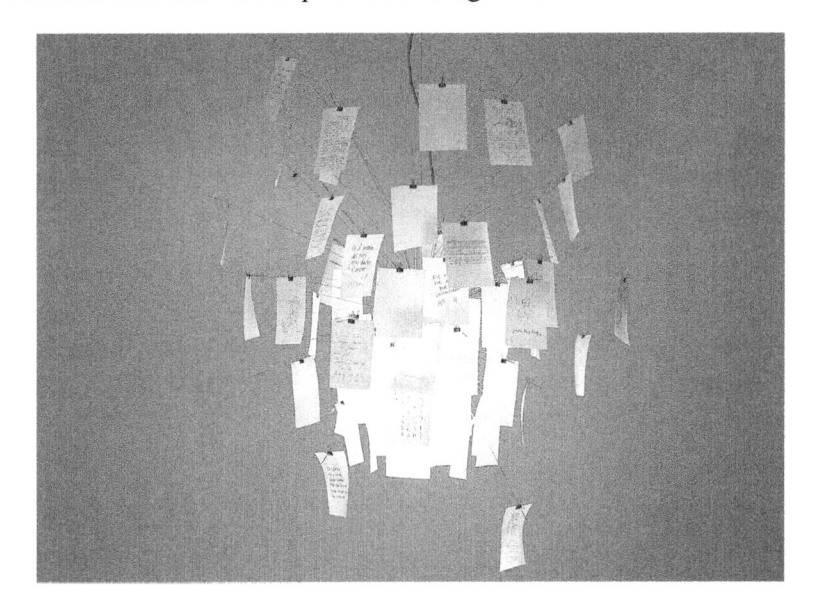

Figure 188: Ingo Maurer's Zettel'z 6 light fitting.

Commonly there is no light intensity distribution curve for these luminaires. If one needs to evaluate their effect on a lighting calculation program, the designer must try and find a similar luminaire with photometric data available and use it instead.

Sometimes these fittings actually become tasks, because they might need to be illuminated by other light fittings to look their best.

For all of them it is very important to apply the usual checks on whether they ruin visual comfort and uniformity.

21.4 Self evaluation

Answer the following questions to test your understanding of the material:

1. Describe traditional asymmetric luminaires: their light intensity distribution, typical light sources, behaviour, ideal tasks, examples of luminaires, examples of systems and the type of illuminance these generate. Apply this luminaire to a real sample interior and discuss the results.

2. Describe wall-washer narrow beam luminaires: their light intensity distribution, typical light sources, behaviour, ideal tasks, examples of luminaires, examples of systems and the type of illuminance these generate. Apply this luminaire to a real sample interior and discuss the results.

3. Describe wall-washer narrow beam luminaires: their light intensity distribution, typical light sources, behaviour, ideal tasks, examples of luminaires, examples of systems and the type of illuminance these generate. Apply this luminaire to a real sample interior and discuss the results.

4. How is a fully asymmetric luminaire a task?

Selecting the fitting by other criteria 22

I am going to group together a set of minor criteria that can lead the designer to the right choice of fitting. These methods are surely as effective as the major ones, however they are less complex and mostly refer to concept already presented.

22.1 Selecting by lamp

Traditionally, selecting fixtures by this criterion means looking for a luminaire that is either miniaturised, has excellent colour rendering, or generates coloured light.

When the first low tension halogen lamps appeared in the seventies there was a revolution in the design of luminaires, because halogens were new, long lasting (for the times) and reasonably powerful. They also had excellent colour rendering and warm colour temperature. Moreover, the limited size of the lamp allowed designers to create a new wave of products (table lamps), with a much more modern look and more flexibility than ever before.

Something similar happened when, in the past few years, LEDs became a viable option for lighting, and subsequently flooded the market.

22.1.1 Colour rendering

Certain tasks require very good colour rendering: enjoying a painting in a museum, choosing and applying the correct make-up, selecting a suite in a store, working in quality control etc.

The most commonly adopted solution was the halogen lamp. Unfortunately its well-known disadvantages (short life, low efficiency, fixed colour temperature, heat) make it a niche product, especially considering the relevant improvements in light quality made by fluorescents, metal halides and LEDs.

22.1.2 Colour temperature

The designer's concept of the interior must contain details on the colour temperature for the lighting. One might select warm light to give a cosier feeling to the interior, and emphasize warm colours, dark wood and leather. Another might prefer cool light to give a more business-like feeling to the room, emphasize cooler colours, glass and steel. At times it can be useful to play with the contrast between warm and cool light in an interior, maybe to create artificial shades or more particular effects.

Most types of lamps have a version that emits warm light, and one that emits cool light. Fluorescent lamps, in particular, come in a very wide variety of colour temperatures, and are the most flexible solution, while halogens can only produce warm light.

For very special applications, such as stage lighting, one can employ a range of colour correction filters mounted near the opening of the fixture, which changes the colour temperature of light.

22.1.3 Size requirements

With the advent of white LEDs powerful and efficient enough to be used in lighting, two new avenues opened up: one related to colour (see 22.1.6) and the other to size. LEDs are by far the smaller light source available, so they made it easier to, for example, have light coming from niches in the ceiling, recessed fixtures, or grazing light effect in the interior. Other notable applications are integrated fittings: cabinet lighting or shelf lighting in their furniture.

When low-tension halogen lamps with dichroic reflectors were introduced, something similar happened, suddenly every interior was illuminated by rows of recessed dichroic downlights, or by track mounted dichroic spotlights that made creative use of the new technology.

22.1.4 Instant start

Some applications require lamps to instantly start (e.g. security systems or motion sensor switches). This means no metal halide lamps, and no traditional ballasts for fluorescents.

22.1.5 Dimming

Some buildings, usually offices, have a centralised light control system that dims the fixtures. This is less of a problem than it was in the past, as now it is feasible to dim, at least partially, not only halogen, but also fluorescents and metal halide lamps. Still, if the requirement is a full 0 to 100% dimming capability, then discharge lamps are out, and the only solutions left are LEDs and halogens.

The best performers in dimming are LEDs because of the way the light is produced by them. A LED will dim its luminous flux without changing the spectrum emitted: the colour rendering and colour temperature of the light will remain exactly the same.

22.1.6 Coloured light

An interior designer might need the light to change dynamically from one colour to another, or might want to obtain some other special effect. Before the advent of LEDs this was done in a very cumbersome way, by using monochromatic fluorescent lamps and a dimming system. Now the same result can be obtained via a RGB LED system, and it is also simple to program very complex colour effects and transitions with (free) software to control the drivers.

22.2 Selecting by luminance control

In the last thirty-five years the use of computers in offices has gone from relatively rare to commonplace. More recently, most screen manufacturers decided to lose the matte film that was applied in front of the LCD screen to get a better colour contrast. Unfortunately this increases the luminance contrasts reflected on the computer screen dramatically.

Tablets are becoming common in the workplace, and their screens are in a new position that is almost guaranteed to receive unwanted luminance.

So, it is now even more important to make sure that the fixtures are compliant to the regulations. In Europe the standard dealing with this is the EN12464-1. Among other things, it sets the maximum allowed luminance at or beyond a certain angle.

The designer must also consider the impact that an area of high luminance left unscreened (e.g. windows, even reflected on a shiny surface such as a glass partition) would have on a shiny display screen in the whole lighting system. Each and every fixed computer workstation needs to be analysed to make sure that in the area reflected by the monitor there are no sources of high luminance contrast.

High contrasts in the lighting system are not automatically a bad decision, because it is the only way to produce higher contrasts on the visual tasks. There is a trade off between high contrasts on the task and glare that needs to be addressed on each installation.

22.3 Selecting by efficacy - Specific Connected Load

The idea is to select the fixture with the highest total efficacy (4.4 on page 43) among those that are fit to illuminate the task.

A single criterion is never enough to make a sensible decision. Let's consider an example: the fixture with the highest total efficacy is… just a very efficient lamp and no luminaire ($R_{LO} = 1$) because anything else will absorb some flux. So the most efficient solution would be to forget about the fixtures entirely and just put naked lamps in our interiors.

But naked lamps mean:

1. No illuminance uniformity, ever;

Selecting the fitting by other criteria

2. No visual comfort, ever;
3. No artificial light in: bathrooms, other humid interiors,
4. No artificial light outdoors, and so on.

I would like to offer a couple of more realistic examples.

1. A very high efficacy, indirect fixture (see 18.3 on page 200) will always waste more energy illuminating a desk than an average direct fixture. The light coming out of the indirect fixture must be bounced back onto the desk via ceiling and walls, which means that the reflectance, the colours, materials, and finishing of the ceiling and walls, the shape of the room, of the furniture and so on, will each affect the amount of light actually reflected onto the task. On the other hand the direct fixture (see 18.2 on page 198) will hit the desk with a much larger percentage of the flux emitted.

2. A very efficient spotlight that has too wide a beam for the task will end up being less efficient than an average spotlight whose beam is just right, because a large part of the flux will be wasted outside the task.

3. The photometric curve and the light output ratio should always be evaluated together: a fixture with an excellent photometric curve might get its results by sacrificing efficiency, while an efficient fixture might get its high R_{LO} by offering very little light control.

A good indicator of the true efficacy of the interior, with its colours, materials, furniture, finishing and its lighting system must take into account both the average illuminance on the task and the power installed every square metre.

I will introduce the Specific Connected Load:

> **Specific Connected Load (SCL):** is the amount of electrical power needed to reach 100 lx on a square metre of the interior. It is measured in watt over square metre over 100 lx. (60)

The larger the SCL, the less efficient the system.

The two main areas where SCL can be useful are:

1. To compare different solutions on the same interior and see which is more efficient.
2. To give the designer an efficacy value to aim for in the lighting system.

For example, the system in 18.2 on page 198 (the conference room illuminated with direct only fittings) has a SCL of 2.21 W/m²/100 lx, while the same room lit by indirect only fittings (18.3 on page 200) has a specific connected load of 25.91 W/m²/100 lx; this means that it needs more than eleven times the electrical power to provide the same 100 lux of illuminance. This indicator easily shows how direct luminaires are inherently more efficient than indirect ones.

This indicator can be calculated by DIALux.

SCL can be applied to evaluating general and task lighting. When designing the latter, it is meaningful to consider the task lighting luminaires only, and compare different system applied to the same task area.

22.4 Selecting a luminaire by beam spread

In a spotlight, or in general in a fixture that has an axially symmetric photometric solid, beam spread is a very synthetic indicator of how wide the usable cone of light is (Figure 189).

Selecting the fitting by other criteria 265

This is especially useful, because the process of dimensioning spotlights starts with two known quantities: the size of the task to be illuminated and the distance between the task and the mounting point. By using a simple trigonometric calculation or the cone diagram, the designer can obtain the needed beam spread (see 5.4 on page 56).

The limit of this method is that, like the R_{LO}, it is a very synthetic indicator: it does not tell anything about illuminance uniformity on the task plane. This is not such a big problem for narrow beam spotlights, which are the fixtures that traditionally use the beam spread, but it becomes one for wide beam spotlights.

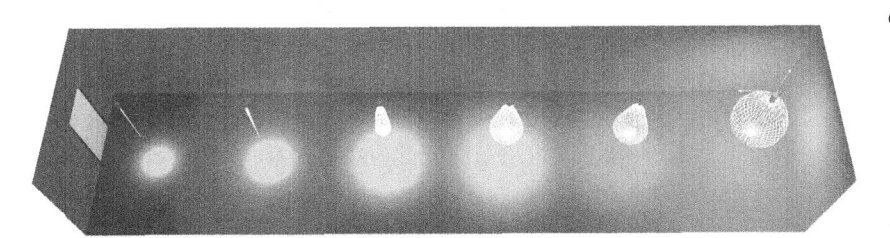

Figure 189: Example of different beam spreads.

Once again this quantity is best used together with the light intensity distribution curve of the fixture.

Please remember that the visual extension of the lit spot on the floor depends on background luminance: if it is low, then the radius of the spot will appear bigger, if it is high it will appear smaller.

A normal spotlight does not trace a definite oval of light on its task: the contours of the bright areas are blurred and the illuminance gradually goes down to zero. In order to get a very definite shadow from a spotlight, one needs to use a system of lenses and baffles that give shape to the light, just like the limelights in a theatrical production.

22.5 Selecting a luminaire by mounting

One of the first decisions is defining the mounting points for the luminaires. In many cases the reader needs to find a way to satisfy two types of constraints:

1. One comes from the interior itself:
    1.1. Previously fitted mounting points.
    1.2. Areas where no mounting is possible.
    1.3. Windows, skylights and so on.
2. Another comes from the necessities of the system formed by the task, the luminaire and their relative positions:
    2.1. For example: if the task is a vertical surface and the position for the fixture is on the ceiling, we know we need a wall-washer, but wall-washers need to be at a certain distance from the task to give their best performance, and we need to make sure that distance exists.

Quite often the mounting positions for a system are fixed or hard to change, so it makes sense for manufacturers to organise their catalogues according to this criterion.

22.5.1 Ceiling mounting

Usually the same fixture can be mounted on the surface of a wall or of a ceiling (Figure 190).

This of course does not mean that it is wise to do so: a ceiling mounted luminaire is almost completely out of view, so we don't have to worry too much about luminance contrast issues there; a wall mounted one, on the other hand, is an entirely different affair.

Selecting the fitting by other criteria 267

Figure 190: Ceiling mounted luminaire.

22.5.2 Wall mounting

Very often this kind of fitting damages visual comfort, being in the middle of the visual field of any person, either sitting or standing, that is looking towards them (Figure 191).

It is then very important that the reader checks against excessive contrasts of luminance caused by either the luminaire itself, or by the reflection of light on the wall around the mounting position of the fitting.

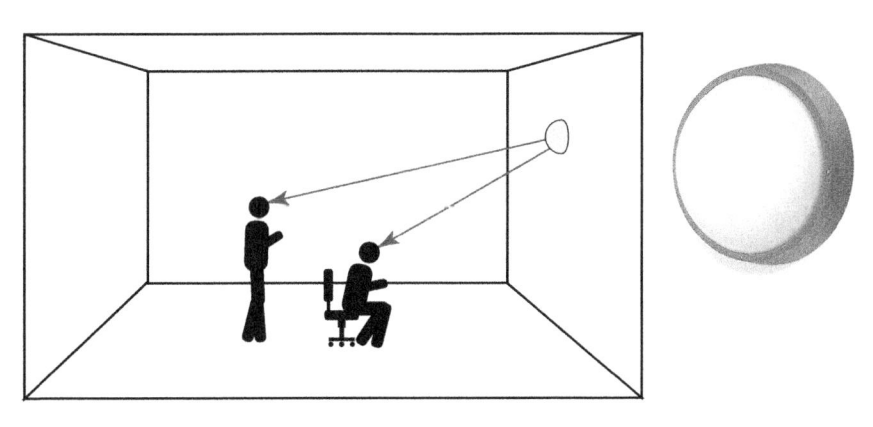

Figure 191: Example of wall mounted luminaire and its issues with glare.

Figure 192 shows in red the directions where high luminance is most damaging: it can easily reach the eyes of an observer, either directly or indirectly (via a smooth ceiling or wall partition), and in yellow the safer ones.

Figure 192: Dangerous directions for glare on a wall mounted light fitting.

Because visual comfort is at risk with this kind of fitting, it is important to verify the screening of the luminaire as well, especially because there are many examples of ineffective ones.

Unfortunately luminance screening means a reduction in efficacy and a light intensity distribution that is usually not very good.

22.5.3 Suspended mounting

Figure 193: Examples of suspended fittings.

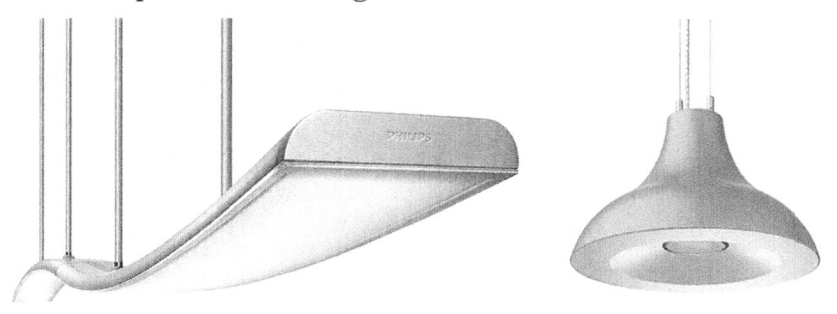

Suspended fittings (Figure 193) are usually in the middle of the field of view of any user of the interior. It is paramount to make the usual checks to protect visual comfort.

The direct-indirect CIE class (see 18.7 on page 208) is especially suited to this class of mounting, because of its luminance control at gammas close to 90°, while general-diffuse luminaires are very often mounted suspended (see 18.6 on page 206), even if they are not particularly apt to it.

22.5.4 Portable mounting and free standing

The obvious advantage here is to allow the user to place the fitting where he needs it. This is not always a good thing: it is the designer's task to create a system that works, the user should not have to change the placement of the fittings in order to enjoy the interior.

Figure 194: Examples of table mounted (left) and free standing (right) luminaires.

The table lamp requires the same checks on its luminance contrast of the wall lamp, with a difference: the area over the lamp is now at risk as well, because it can dazzle the users that walk too close to a low one.

Floor mounted fittings usually create "islands" of brightness in an interior. They are commonly associated with a sitting device of some kind, and with a system with lower general illuminance.

Some floor-mounted fittings are used to provide general illuminance in an interior without ruining the lines of the walls and ceiling. Unfortunately this can glare on users, either directly by the fitting, or indirectly by luminance on the walls. The reason for this is that the luminaire should be, at the same time, as low as possible, to cover much of the ceiling, and as high as possible to be out of the field of view.

Moreover it can be difficult to obtain illuminance uniformity, because of the limited field illuminated by this type of fitting.

22.5.5 Recessed mounting

Recessed fittings (Figure 195) have the advantage of completely disappearing from view. The obvious disadvantages are the impossibility of moving the fixture without a lot of work, and the need for a false ceiling.

Figure 195: Examples of recessed luminaires.

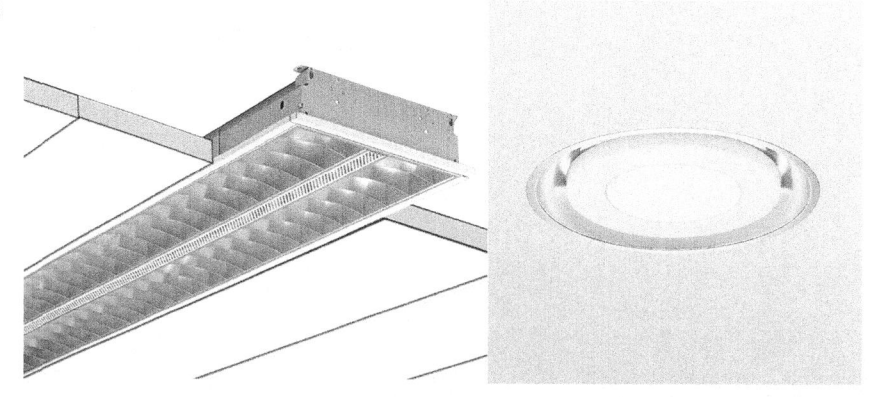

There is a small number of floor-recessed fittings, but their use in indoor lighting is extremely rare, because it is too easy to glare up a person that walks close to them.

A few wall-recessed fittings exist as well: they are usually employed to light the floor or the ceiling. They tend to cause glare as well.

22.5.6 Track mounting

Much more flexible, track mounted fittings (Figure 196) are very often employed in commercial areas and museums, because they make it very easy to adapt the existing lighting system to a new disposition of the merchandise or the exhibits.

The most common type of track-mounted fixture is surely the accent light, in all its declinations (narrow beam axial-symmetric or wall-washer, and so on). This type of fitting is of course very prone to glare by bad aiming.

Figure 196: Examples of track mounted light fittings.

22.6 Selecting a light fitting by commercial classification

To make it easier to see one's way through the huge number of fittings available on the market, it is, in my opinion, very useful to spend time looking into the commercial classification of fittings.

Unfortunately this is often confusing. For example, commercial wall-washers do not always have a wall-washer light intensity distribution (17.4.2 on page 264), they can simply be an asymmetric, or even just tilted spotlights. Moreover, the same fitting can belong to multiple categories (downlights and troffers, industrial and battens).

22.6.1 Downlight

Downlights (Figure 197) are fixtures usually recessed into ceilings and have a roundish shape.

Their light intensity distribution is axially symmetric or symmetric around two planes with a medium to wide beam spread, and they can have trims or other adjustable parts to aim their light.

Usually these types of fixtures provide general lighting for hospitality, work and commercial areas.

The lamps used are halogen, compact fluorescent or metal halides.

Figure 197: Examples of downlight.

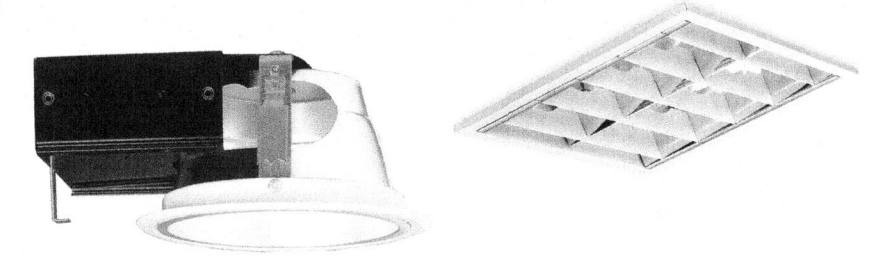

The choice of halogen lamps today is almost impossible to defend: they do have the best colour rendering, but their very low efficacy is too strong a disadvantage in general lighting, where it is necessary to produce large quantities of luminous flux.

22.6.2 Troffer

It is usually a fluorescent fitting, but there are LED versions as well (Figure 198), shaped like a square or rectangular box containing lamps, reflectors and wiring.

Figure 198: Examples of troffer.

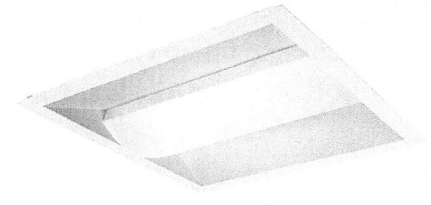

Most of these fittings are recessed, like the one shown in Figure 198, usually into a tiled ceiling, so their size is either 60 cm by 60 cm (2' by 2') or 60 cm by 120 cm (2' by 4').

A few of these luminaires are surface mounted, generally on the ceiling, like the one in Figure 199.

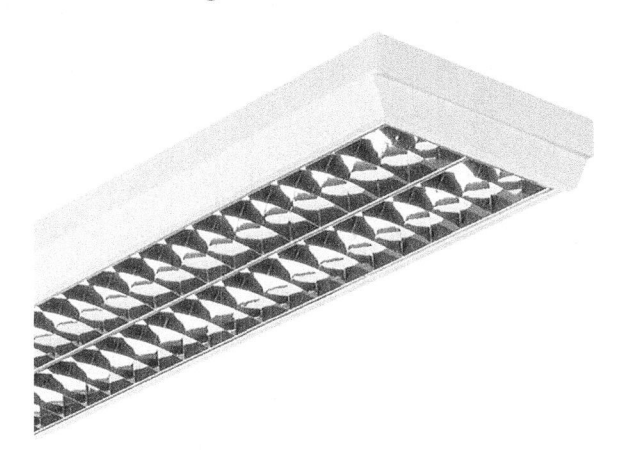

Figure 199: Picture of a surface mounted troffer with anti glare screen.

22.6.3 Commercial fluorescent

They are often the cheapest versions of fluorescent fittings. The most common subtypes follow.

22.6.3.1 *Wraparound diffuser*

Unsurprisingly, these fitting have a diffuser (Figure 200) that can be either clear or frosted. It has three functions:

1. Protection against humidity and dust: these fittings can have high IP ratings.
2. Luminance contrast reduction: the fitting can be in the field of view without damaging visual comfort.
3. Emphasising the diffused light effect one obtains from fluorescents.

Figure 200: Examples of wraparound diffuser.

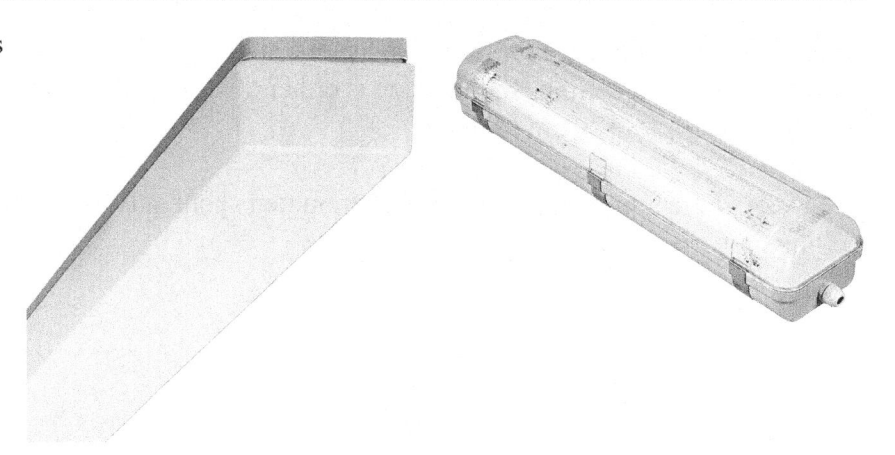

22.6.3.2 Fluorescent strip - batten

It is the most basic of fixtures (Figure 201): just a holder for the lamp that houses ballast and starter. It can only be used in a non-aggressive environment.

The main strengths of this type of fixture are its price, the cheapest possible, and its size, as it's not much bigger than the fluorescent itself.

Figure 201: Examples of batten.

It can fit into small recesses, or it can be hidden behind architectural features or screens. Before LEDs, it was the best way to hide a light source in a small space and provide enough luminous flux for general lighting.

22.6.4 Industrial fixtures

Historically, these fixtures were designed to provide general lighting for workshops, factories, warehouses and so on. They needed to be robust, efficient and easy to maintain.

They are normally used in interiors, but there are special versions with higher IP ratings that can face a much harsher environment. Now they are a common sight in industrial styled interiors as well.

Originally, these fixtures used efficient lamps with poor light quality (e.g. mercury-vapour lamps). When using them in non-industrial interiors, the designer must select a more modern and high quality light source, such as metal halides or fluorescents.

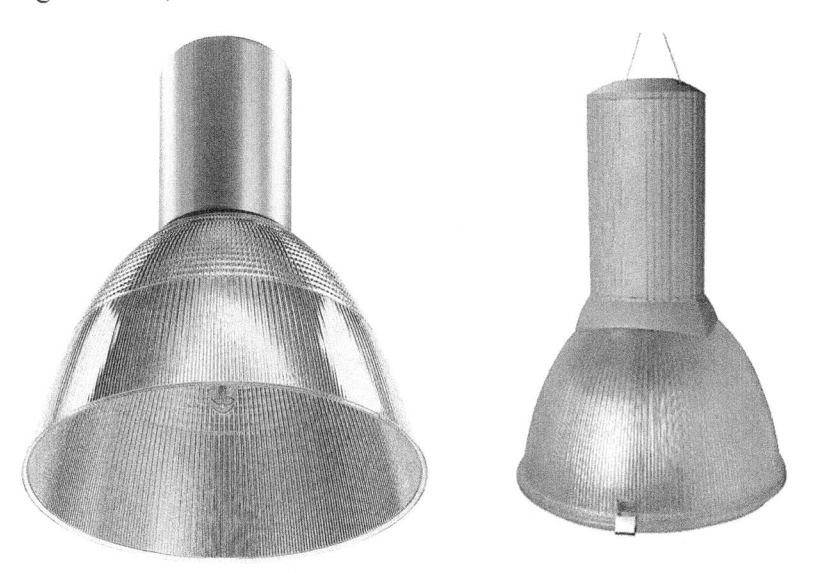

Figure 202: Examples of industrial light fittings.

Fluorescent fixtures are very common, because of the high performance of that type of light source, and they vary from a simple holder for the lamps and white reflector (Figure 201), to waterproof fixtures for outdoor use (Figure 200). There are suspended versions of them (Figure 202) which have metal halides or compact fluorescent lamps.

22.6.5 Linear systems

These types of fixtures have been illuminating offices for the last 30 years. They often are suspended from ceilings and placed over the desks.

They try to keep the same overall shape, usually very technical, while combining different functions, each in a simple module that can be connected to the rest of the line, or just placed close to it.

Figure 203: Examples of linear system.

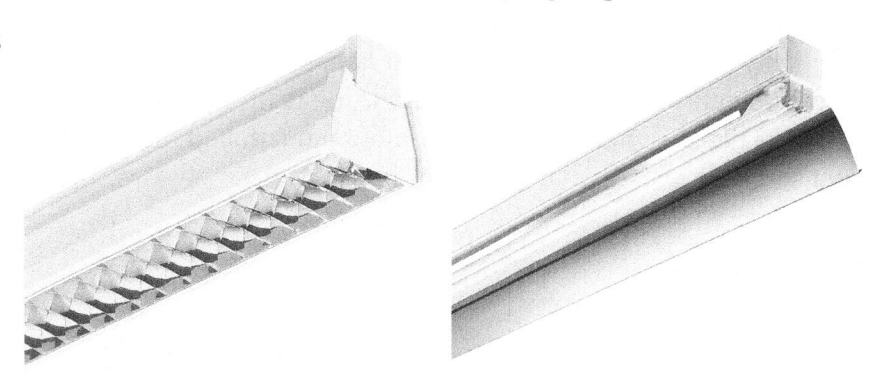

The modules usually are: direct, diffused, with luminance control, with spotlights for accent lighting, or indirect.

22.6.6 Wall-washers (commercial definition)

Wall-washers are fixtures usually mounted on ceilings; their tasks are extended in relative size. These fixtures can sometimes be recessed. So they illuminate anything vertical (a bookcase, a wardrobe, etc.) while being mounted on a horizontal surface (or vice versa), with a good uniformity and without being in the field of view of the user.

The most common shapes are boxlike or cylindrical.

We know that a wall-washer fixture might not have a wall-washer light intensity distribution, so it is important that the designer checks.

Selecting the fitting by other criteria 277

Figure 204: Example of eyelid wall-washer.

This type of fitting serves as a reminder of definition (62) on page 257, i.e. that the selection of the right luminaire for the task comes from studying its light intensity distribution, not its shape.

22.6.7 Wall grazer

This fixture is usually mounted on the ceiling next to the walls, often recessed.

The wall grazer's nature is that of a theatrical effect, or, from a more technical point of view, it is illuminating a vertical task with horizontal illuminance.

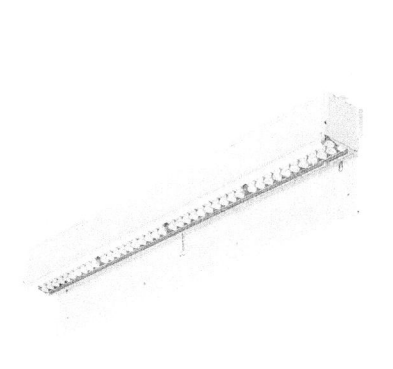

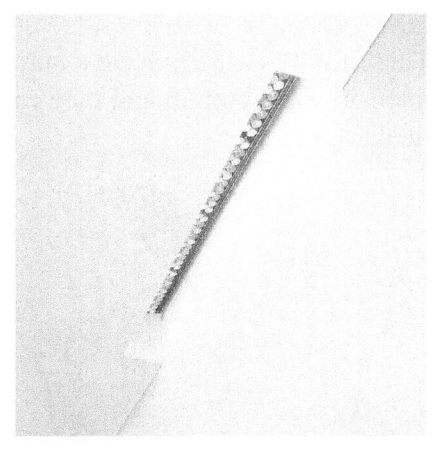

Figure 205: Example of wall grazer.

It is the most extreme way of highlighting the surface's roughness, because any minute irregularity will cast a pattern of light and shadow when the light is almost parallel to it.

Visually it is the opposite of the effect one obtains with a wall-washer, which will tend to flatten the surface because light comes from a direction closer to its normal.

The fixture needs to have (ideally) a very narrow beam crosswise from the wall. Lengthwise from the wall, it depends on the fitting:

1. A single light source one with a very narrow beam will draw a column of high luminance on the wall.
2. A wider beam luminaire will draw triangles growing wider as the beam widens.
3. A multiple light source luminaire, such as the ones in Figure 205 will brighten up the whole width of the wall.

The real issue here is to be able to reach the bottom of the wall with the beam of light without creating pools of very high luminance on the floor right next to it.

22.6.8 Accent fixtures

These fittings usually have a narrow beamed, axial symmetric light intensity distribution; and they can be aimed at and highlight a particular task.

Figure 206: Examples of accent light fittings.

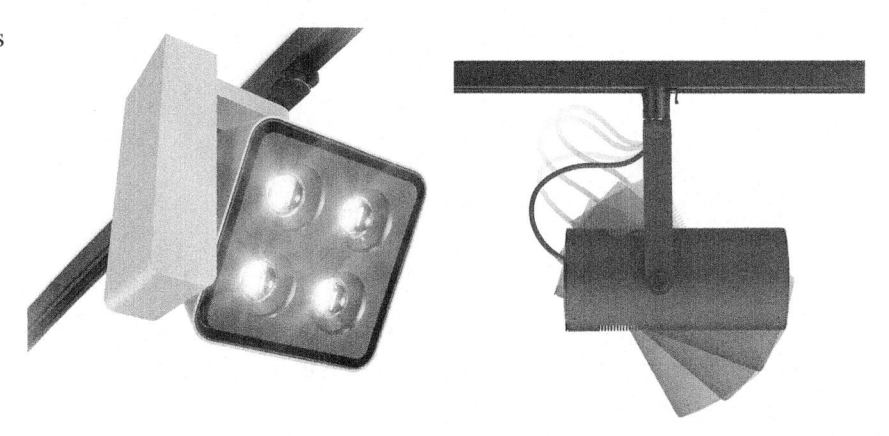

They can be recessed, or mounted on walls, ceilings or, more commonly, on a track.

They are flexible, especially when track mounted, because the user can easily change the aim and position of the fittings, thus accommodating a large number of tasks. This commonly happens when a store, a museum and so on need to rearrange their furniture, merchandise and exhibits.

It is a good fixture to associate to the downlight in commercial interiors, because one can provide general lighting and the other task lighting.

22.6.9 Cove lights

These fixtures are mounted recessed or hidden into coves and are used to highlight architectural features in an interior.

The need for small fixtures has compelled the designer to choose thin fluorescent tubes held up by very small lamp-holders, until LEDs became powerful enough to be used in lighting.

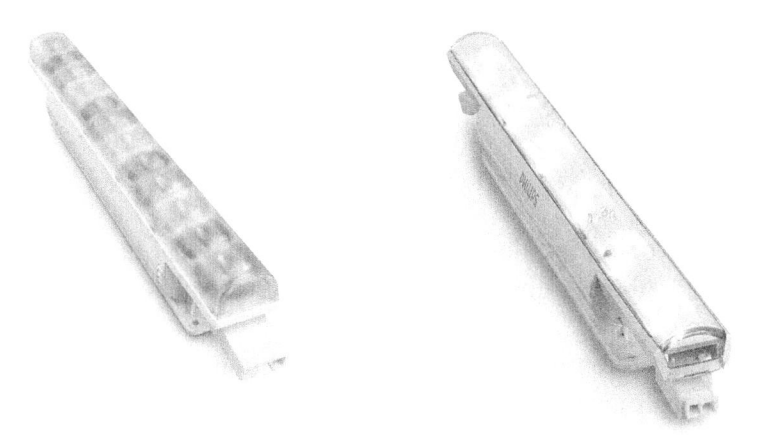

Figure 207: Examples of cove lighting: RGB LED (on the left), and white LED (on the right).

LEDs gave these fixtures much more flexibility, so that they could be employed in smaller coves; improved their performance, as there is now no low luminance spot caused by the area between the end of one tube and the beginning of the next; and gave the designer the ability to play with colours when RGB LEDs were introduced.

It has become very easy to program complex sequences of changing colours in the interior, so that a designer could invent new, creative, but not always sensible special effects.

22.6.10 Decorative lights

Many different types of fixture belong to this category: floor lamps, pendants, chandeliers, drum luminaires, table lamps, sconces, lanterns and so on. All these different fixtures share a low technical content: very often they don't have a photometric curve capable of illuminance uniformity. The primary concern in designing these fixtures has been their look, not their performance.

The designer must take particular care when employing fixtures like these: since they are unable to provide uniform illuminance they must be integrated with fixtures that can, maybe hiding the more technical fittings into recesses, so as not to mar the look of the interior.

Some of these fixtures, wall sconces for example, are inherently prone to causing glare, especially around $\gamma=90°$.

22.6.11 Task lights

It's a type of fixture designed to illuminate a local task. Examples are under-cabinet lights for kitchens or vanity lights for mirrors.

Mirrors are often badly solved by the lighting system, because designers and manufacturers seem to forget that the task is the person looking into it. Many bathroom mirrors have luminaires mounted on top of them that provide horizontal illuminance, when the person needs semi-cylindrical illuminance to be well lit.

The second objective with mirrors is to have relatively low luminance contrast in the field of view of the observer.

The system in Figure 208 is a good solution of the problem.

Figure 208: A mirror with an integrated low luminance lighting system that generates semi-cylindrical illuminance on the person.

22.7 Self evaluation

Answer the following questions to test your understanding of the material:

1. Explain the reasons why one might need to select a luminaire by its lamp.
2. Explain the selection by luminance control.
3. Describe in detail the concept of Specific Power Load.
4. Give an example on how to find the correct beam spread for an installation.
5. Describe each different way of mounting a luminaire, with its advantages and disadvantages.
6. What are the issues with wall mounting?
7. Explain the advantages of track mounting.
8. Describe each commercial type of luminaire.
9. Invent a different solution to illuminate a mirror that satisfies all the requirements.

Part 4

This part describes the work-flow necessary to create a lighting system. It is a task that requires two different viewpoint: the purely technical one (it will make things work) and the purely decorative one (it will make things look good).

The good lighting designer is supposed to create beauty that functions. In order to do that, he needs to be aware of many aspects of the interior where he plans to work. He has to gain an intimate familiarity with it, in order to recognise all its requirements and its necessities, and prepare a system that will meet them. It is not only a matter of knowing the interior well, having visited the location (whenever possible) and having studied the plans; it is also a matter of knowing the activities performed there, and how and where people will use the interior and perform the activities.

In short, the designer must be familiar with:

1. The location.
2. The activities.
3. The people.

But the designer needs familiarity with the concept as well. Even when it is fully visualised or rendered, the actual behaviour of light in the interior may not be really clear to the designer.

The concept

When the designer imagines the concept for his interior, he defines implicitly much of the lighting system, because he decides, among other things:

1. What is there to see?
2. What will be highlighted?
3. From where should the light come?
4. What kind of light to use.

Unfortunately, many of these decisions are done unconsciously: the designer is in a certain sense too used to light and not at all familiar with it. It is not considered an inherent property of every component of the interior, just something that is somehow there. Where it comes from and how it got there are questions that remain unasked.

It takes time and reflection to answer these questions, even if the designer is perfectly able to visualise the interior as he imagines it.

There is a rush to producing a light plan that should magically solve all the problems, but it often turns out to be completely disconnected both from the concept and from the actual technical requirements of the interior.

The process between light in the concept and actual lighting system is completely missing, because the designer doesn't have either the tools or the awareness required to complete it.

23.1 What is there to see?

Let's start tackling the first question. The objective here is to gain a conscious knowledge of the types of visual items and of the properties of their appearance, using an interesting model[1] that explains how visual perception of an object works, and how to transform the different modes of perception into a real lighting system.

The model identifies several types of items and their attributes that can be perceived. The link is very strong: only a certain category of visual task will display certain attributes, and vice versa (Table 30).

Table 30: Modes of appearance and attributes.

Object	Attribute	brightness	lightness	hue	saturation	flicker	pattern	texture	gloss	clearness
located	illuminant	●		●	●	●	●			
	illuminated	●		●	●	●	●			
	object surface		●	●	●			●	●	
	object volume		●	●	●			●		●
non located	illuminant	●		●	●	●				
	illuminated	●		●	●	●				

The model distinguishes two main categories of items: localised and non-localised.

Located items are those limited by a surface, such as the bright surface of a light fitting or a patch of sunlight on a floor. There are three modes of located appearance:

1. Illuminant mode

It is what we perceive when we look at a lamp or a light fitting: something that is self lit.

---

1. Adapted from CUTTLE, MODES OF APPEARANCE AND PERCEIVED ATTRIBUTES IN ARCHITECTURAL LIGHTING DESIGN

2. Illuminated mode

It is the effect of a patch of sunlight on the floor: a surface illuminated by something else

3. Object mode

There are two modes in which we perceive objects:

- 3.1. Surface mode: any opaque object seen because of the light reflected on its surface
- 3.2. Volume mode: a translucent object, smoke or a cloud.

Non located items do not have a boundary surface. They are:

1. Illuminant

Illuminants generate light and do not have a boundary. Examples are the sky, or ambient fog (strictly speaking fog does not generate light, but the perception of it is the same).

2. Illuminated

The perception of general lighting inside a room.

23.1.1 Attributes

Each task exhibits certain attributes and not others. The list of attributes, and their relation to the items, is shown in Table 30. Let's look at each attribute and find out more.

Brightness is a property of the light, while **lightness** is a property of the colour of the surface. It is interesting that they are mutually exclusive, and it is what allows artists to give the impression of three dimensions using only colours on a flat canvas: the viewer is tricked into believing that the circle painted white to black in an appropriate manner is actually a sphere in three dimensions illuminated from the side.

Hue and **saturation** are the two remaining properties of a colour, and, as such, are perceivable everywhere.

Flicker is a property of light, so it is not present in objects.

Pattern requires a located mode of perception.

Texture and **gloss** are properties of a surface, and **clearness** of a translucent volume. As such they appear only on object modes.

Figure 209: Application of the Cuttle model to a diffused luminaire.

Writing down and studying in detail where these different modes appear in the interior and describing their attributes helps the designer formalise his concept of the interior and implement his vision.

23.1.2 Example of use

Let's apply the table to a **located illuminant**: a diffused luminaire (Figure 209 and Table 31). **Hue** and **saturation** will always be visible, because they describe colour. **Brightness** gives us a qualitative indication of the luminance of the luminaire, **flicker** tells us whether the luminance is steady in time or not, and **pattern** describes the local characteristics of the surface.

The concept

Being an illuminant, **lightness** is replaced by brightness, the **texture** of the surface, the **clearness** and the **gloss** are not perceivable because the luminaire is emitting light.

Attributes	Diffused luminaire	Fog
Hue and Saturation	White	Grey
Brightness	High	Low
Flicker	Stable	Stable
Pattern	Flat	not perceived

Table 31: Cuttle model result for a diffused luminaire and the fog.

Let's apply the table to a non located illuminant this time: fog.

As usual, **hue** and **saturation** will be visible, and determine the grey colour of the fog. **Brightness** again is perceivable instead of **lightness**, because fog is luminous, and its steadiness is seen as **flicker**. There is no **pattern** because the fog is not located, nor there is **texture** or **gloss**, because the fog has no surface and is emitting light. Finally there is no **clearness**, because fog has no volume.

What will be highlighted?

This is an important question in the design of the interior and lighting system. The designer must ask himself what it is that he wants to show in the interior, and create the lighting accordingly.

23.1.3 Luminance and surfaces

The effect of selecting, for example, what wall will be lit, or whether to leave the ceiling dark or not, will significantly influence the ambiance and the perceived shape of the interior.

Let's look at a few examples in the following sketches.

On the left side of Figure 210, the illuminated back wall appears farther away; on the right side it is dark, and it appears nearer, making the interior appear less deep than it is.

Figure 210: Simple sketch of an interior with bright walls (left) and a dark back wall (right).

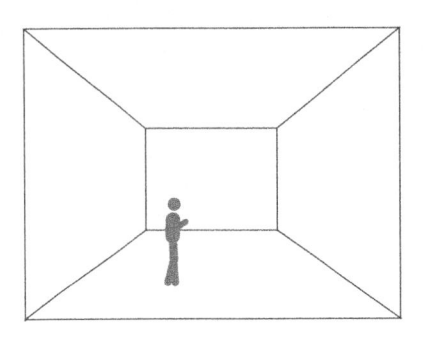

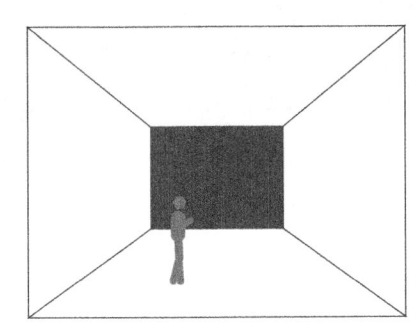

On the left side of Figure 211 on page 290, there is a sketch of an interior lit with constant luminance: it looks rather flat and the depth perception is reduced. On the right, the exact same interior with the walls left dark: the interior appears narrower and there is now a better perception of depth.

Figure 211: Simple sketch of an interior with bright (left) and dark (right) walls.

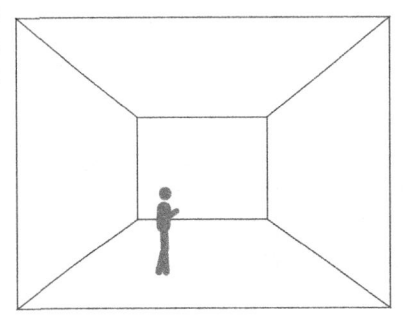

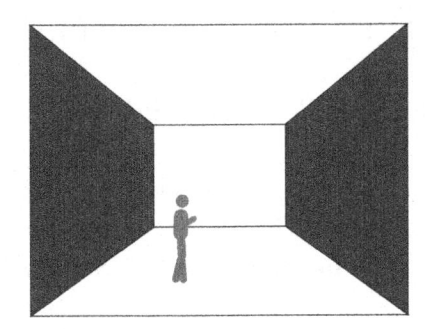

The right side of Figure 212 shows the effect of wall grazing: it makes the ceiling appear higher and less oppressive.

The concept

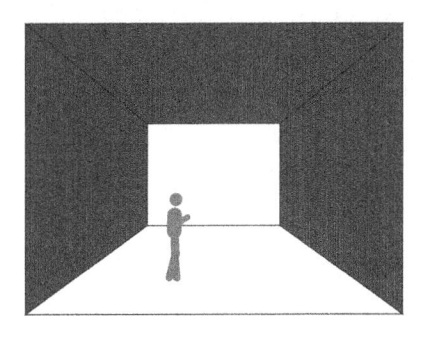

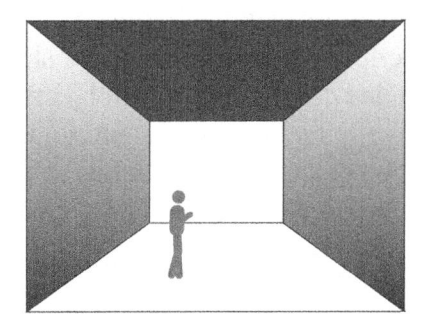

Figure 212: S i m p l e sketch of an interior with a dark ceiling and dark walls (left), or wall-grazing lighting (right).

Figure 213 shows the effect of low luminance on a surface: on the left the dark walls make the room appear narrower, on the right the dark ceiling appears lower.

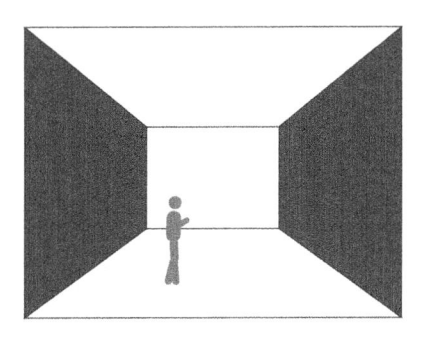

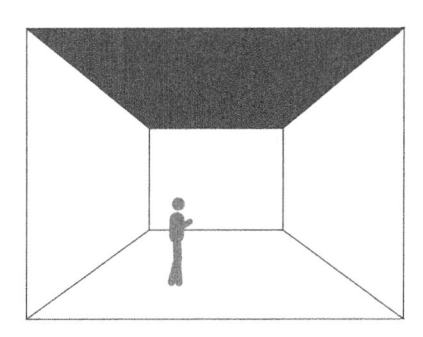

Figure 213: S i m p l e sketch of an interior with dark walls (left), or a dark ceiling (right).

The simplest way to implement these choices is to select the correct luminaire according to the CIE method (see 18 on page 197).

23.1.4 Photo realistic rendering

It is a very important tool to visualise and design the interior, and its ability to produce excellent results cheaply has increased with the improvement in technology and software.

It contains a detailed lighting plan, which usually tempts the inexperienced designer to replicate it in the real lighting system. This obviously does not work, for a very simple reason: its light sources are not built to be realistic in their behaviour, they need to be realistic in their result. It would be akin to expecting to actually live in a film studio reconstruction of an apartment…

Photo-realistic rendering should be used as an excellent visual description of what the interior should look like, some sort of ultra-detailed description of the objective. In order to implement this objective in reality, the designer must apply the tool set he derives from the book to build the actual system and the interior.

23.2 From where should the light come?

The uniformity on the visual tasks is not the end of the lighting system. A person moving in the interior will be illuminated differently according to where he is, where the different light fittings are, and so on. A person standing still and observing the interior, on the other hand, will perceive the flow of illumination, and how the final effect depends on the position, orientation and extension of the objects. A couch lit by a torchères, for example, will have a brighter side, next to the fittings; the brightness will follow the surface of the sofa, and grow dimmer and dimmer as one moves farther away from the luminaire.

These experiences tell us that lighting is actually a three-dimensional field, and its rate of change in space or in time is something that needs to be visualised in the concept.

23.3 Sharpness and flow

There are two main consequences to the observation that light is three-dimensional: the fact that light has sharpness and that it has flow.

> `Sharpness` expresses the capability of the light sources to project harsh shadows and (61)
> very definite pools of light.

The concept

It depends mainly on the apparent extension[2] of the light source as seen by the task: if it is small and point-like, it will provide high sharpness; if it is large, then there will be no sharpness.

> `Flow is the visual indication that there is a stronger light source in a specific direction.` (62)

This will create a brighter area in the direction of the more powerful light source, and a darker area opposite to it. If, on the other hand, there is no stronger light source, the object will be uniformly bright in all directions. Clearly, flow is an indicator of how diffused the light is: the least diffused, the higher the flow.

Figure 214: Example of Cuttle model lit by point-like source. Notice the sharpness on the black ball and on the single shadow, and the flow on the satin white sphere.

There is a model[3] that can help visualise the effect of sharpness and flow. It is made of two spheres, one painted glossy black, the other satin white, and a sundial composed of a circular surface with a rod sticking out of it, here replaced by a candle (Figure 214). Let's examine the behaviour of the sample under three different lights.

---

2. That is, the solid angle of the source as seen by the task.

3. (CUTTLE, LIGHTING BY DESIGN 2008)

23.3.1 Single point-like source

Under this light (Figure 214) we do have sharpness, because there is a small pool of bright light on the black sphere, and a defined shadow cast by the rod.

We also have flow, because the white sphere shows a side brighter than the other.

23.3.2 Multiple point-like sources

In this situation we will still have sharpness because we will see multiple bright areas on the black sphere and multiple defined shadows cast by the rod.

We would normally have flow, because there usually is one source producing substantially more illuminance, or the sources are placed asymmetrically around the task (e.g. all on one side).

Only in the very rare situation in which the sources surround the object equally, and there is substantial equivalence in the illuminance each produces on the object, then the white sphere will be uniformly lit and flow will disappear.

23.3.3 Diffused light from one side

This example (Figure 215) shows low shine, localised on the illuminated (right) side of the model.

We do have flow, as the side of the matte sphere towards the light is much brighter than the other side.

Figure 215: Example of Cuttle model with diffused light coming from the right.
Notice the lack of sharpness and of flow.

23.3.4 Fully diffused light

In this example, shown in Figure 216 on page 296, we have no sharpness: the spot of light on the glossy sphere is rather diffused, and the shadow from the rod almost non-existent.

We have no flow either, as no side of the matte sphere is brighter than the others.

This example represents the other extreme from 23.3.1 on page 294.

23.4 What kind of light should we use?

This entire book tries to provide an answer to that question. In particular Part 1, 2 and 3.

23.5 Self evaluation

Answer the following questions to test your understanding of the material:

1. Describe the visible characteristics of an interior.
2. Explain in detail the Cuttle model, describing the modes of appearance and the attributes.
3. Apply the Cuttle model to a located illuminated, object volume, and non located illuminant.

Figure 216: Example of Cuttle model with fully diffused light.
Notice the lack of sharpness and of flow.

4. Explain what is three dimensional lighting.
5. What is Sharpness?
6. What is Flow?
7. Describe lighting by different numbers and types of sources, and their effect on Sharpness and Flow.
8. Explain how luminance can change the perception of the surfaces of an interior.
9. Invent a lighting system capable of creating the effects shown in each of the pictures in paragraph 23.1.3 on page 289.

The interior

24

The designer must strive towards getting the deepest possible familiarity with the interiors on which he is working; in particular, when dealing with a pre-existent building, it is paramount to:

1. Have a detailed knowledge of the architectural features of the interior, its design, its materials and finishing.
2. Study carefully the activities that are going to be performed.
3. Have at least a partial idea of where in the interior the activities will be performed.
4. Investigate and determines how the users will move in the interior.

In short, the information needed so that the designer can create his own mental image of the interior are those same information needed to design the lighting system. What is missing is the consciousness of how lighting works.

24.1 Architectural features

The accurate evaluation of the architectural features of the interior, and the actual personal inspection of the space are useful for a plethora of reasons.

One can gain a hands-on insight on the volumes, the surfaces, the eventual presence and disposition of glass and transparencies, and their relative extension.

The interior can be inserted in its historical context, and one can decide whether to follow it and select fittings and a type of lighting that fits in with the times, or not. For example, illuminating an Art Deco railway station using suspended fluorescent office light fittings might not be an excellent idea, even if it worked. A Gothic cathedral and a Romanesque one had very different lighting when they were designed. The Gothic style's most recognisable feature is the pointed arch: this makes for taller structures and allows sunlight to enter deeply into the church, while the semi-circular arch of the Romanesque does not. A cleverly designed lighting system should take this behaviour into account.

Other architectural features might be identified and highlighted or integrated into the lighting system design: stairs, arches, niches, features of the ceiling, and so on, might need to be lit, allow for certain special effects (e.g. light grazing), or dictate a certain mounting or type of light fitting.

24.2 Materials and finishing of the surfaces

The main considerations to keep in mind when defining materials and surface finishing for an interior are four: protection, luminance, reflectance and colour bleeding.

24.2.1 Protection

Delicate objects and surfaces, such as paintings, require a balance of very high colour rendering and protection against UV damage. The lighting must take this into account and to do that the designer must identify all the surfaces that need special care in the survey.

24.2.2 Luminance

Selecting glossy surfaces requires careful evaluation to avoid damaging the interior's visual comfort because of excessive luminance contrasts.

Using glossy counter tops on a kitchen, for example, and a high luminance under cabinet lighting system, can easily cause glare. The under cabinet light fittings inevitably end up in the offending-zone (see 9.4.3 on page 112) of the user of the kitchen and cause excessive contrasts of luminance on the work plane. To reduce this prob-

The interior

lem, the solution is to use low luminance fittings and position them so that they are as far out of the offending zone as possible. Or just select a matte counter top.

24.2.3 Reflectance

Very low reflectance on walls and ceiling should suggest that the designer employs direct luminaires, because doing otherwise would penalize the Specific Connected Load of the interior and obtain no real result. The luminous flux sent on the walls will be absorbed, the luminance of the walls will remain low, so they will be hard to see, and they will not reflect flux onto the task.

Any task located on the wall should be illuminated by carefully dimensioned spotlights: they need to highlight the object precisely so as not to waste energy, and avoid excessive luminance contrasts with the background.

24.2.4 Colour bleeding

An interior with extended highly saturated surfaces, especially with a glossy finishing that improves mirror reflection, is prone to colour bleeding. This requires a more careful placement of the luminaires: the designer must consider what happens to coloured light after it is reflected from the saturated surfaces. For example, a glossy red wall next to the tables of a restaurant will not work if lit with a wall washer, but it might if lit by indirect wall or floor mounted luminaires.

Another solution might be to split the system into direct high CRI lighting for the tasks, and a separate set of luminaires for the coloured surfaces.

24.3 Movement of the users in the interior

Knowing how people flow in the interior is very useful for the project and important for the lighting system as well. It helps positioning and aiming luminaires, because when one knows where people are coming from, where they can go and where they will look, one can find the most effective positions for light fittings, and the best aiming points. The objective is to illuminate the task and at the same time to preserve visual comfort for the users in their normal use of the interior.

It also helps defining the types of illuminance required by the task. A statue, for example, when it is in the middle of a room, will require cylindrical illuminance. If it is placed right next to a wall, then semi-cylindrical illuminance will be enough. If it is inside a niche, then one might possibly manage with vertical illuminance only.

24.4 Self evaluation

Answer the following questions to test your understanding of the material:

1. Describe a possible work flow that would bring the designer to become familiar with the interior and its different aspects and requirements.
2. Explain how the properties of the surfaces of the interior affect lighting.

The activities

25

Here I introduce a work flow to make sure all activities are accounted for in the lighting system.

25.1 Interview with the client

The survey of the interior must list the activities performed: what they are, where they are, what mounting points are available for the luminaires, and the positions of any opening, or transparent partition to the outside and between locations inside.

At the same time, an in depth interview with the users of the interior can clarify their necessities and how they perform their activities.

It is important that the designer maintains a positive if sceptical attitude towards the client and the users of the interior during the interviews, because if on one hand, the client is the one that knows best how to perform the activities, on the other hand the client is not an expert in interior or lighting design

When the designer has a good idea of the activities that are performed in the interior, it is a good time to look them up in the literature, to get details (such as the requirements) and to make sure one is up to date with the recent developments.

25.2 The activity table

We know that there is a required level of illuminance, uniformity, glare rating and colour rendering for each activity. Food preparation in the kitchen of a restaurant, for example, requires at least 500 lx of average maintained illuminance, 80 of CRI, 0,60 of uniformity, and unified glare rating under 22.

The project needs to solve each activity performed in the interior.

i.e. satisfy the requirements of every one of them with the lighting system. A good way of going through this is to create an Activity Table for every activity identified in the project. It is a table that contains one activity and its lighting requirements. Continuing with the example of the restaurant, Table 32 contains the requirements for the food preparation activity, and it is incomplete, for now.

Table 32: Activity table with requirements.

	Activity: Food preparation			
Requirements	lx	CRI	U_0	UGR
	500	80	0.6	22
empty for now				
empty for now				

25.3 Identification of the task areas

The next step is the identification of all tasks areas[1] belonging to an activity. A practical way of making sure that all visual tasks have been identified and solved is to update the activity table (see Table 33). The designer needs to add all the task areas on which the activity is performed.

Table 33: Activity table with the task areas added.

	Activity: Food preparation			
Requirements	lx	CRI	U_0	UGR
	500	80	0.6	22
Task area	**Location**			
Kitchen counter	next to wall, under cabinets			
Island counter	free standing in the middle of the room			
...	...			

Some activities can be solved by a fixes set of light fittings. This makes things easier if the activity moves or if the placement of the activity is uncertain in the interior, because the set of luminaires will

---

1. For the definition and details on tasks areas, see 6.2 on page 62

move with it. This happens for example when illuminating a counter top: the fittings, usually mounted under the cabinets, will follow the counter if the blueprint of the interior changes.

25.4 Compliance

The designer will find luminaires and placements that satisfy both technical and visual requirements of the concept of the interior, following the instructions from Part 2 on page 121, Part 3 on page 185, Part 4 on page 283.

Once that is completed, the designer should calculate the results of the project (Part 5 on page 315), and finally check the compliance coming back to the activity table, to make sure that no task area has been forgotten.

25.5 Self evaluation

Answer the following questions to test your understanding of the material:

1. Imagine an interview with a client to illuminate a clothes store. Proceed with the activity table, look up the requirements and indicate all task areas you can think of. The next time you go to a store to buy clothes, check to see whether you have forgotten something.
2. Describe the Activity Table, its function and how it is compiled by the designer.

Layered design

26

Layering the design of the lighting system makes it simpler and more effective, because the whole system is divided into coherent components that are much easier to solve.

The idea is to divide the light fittings because of the function they absolve in the interior: decorative, task or general.

It is important to note that one light fitting can be part of more than one layer, because it can have more than one function.

Together with layering it is important to look into visual task analysis and separation of the different task, so that the problem of implementing the lighting system is brought down to a manageable size.

26.1 Decorative layer

The decorative layer is composed of fittings chosen for their aesthetics, according to the concept of the interior.

Colour effects and RGB LEDs belong to this level as well. For this type of lighting, one needs to make sure that the colour does not bleed into visual tasks that have a colour rendering requirement.

Strictly speaking, the large majority of light fittings will belong to this layer: apart from recessed or other hidden luminaires, the artistic component is an important factor in the choice of a fitting.

Unfortunately, more often than not, the beauty of a fixture is the only function solved by the inexperienced designer. Neither performance nor visual comfort is considered.

This situation is made even more problematic by the lack of performance common to most of the decorative luminaires: very often they illuminate badly and compromise visual comfort.

It is important to conduct both analyses at the same time: the aesthetic and the technical, to find out whether the fitting is just a sculpture or something that can illuminate.

The "light sculpture" kind of fitting is perfectly fine to use: the designer can for example employ them as strong points in the interior. They require two types of checks:

1. They must work.
2. They must not damage the visual comfort.

These fittings work when they are shown to their full effect; this means that they become themselves a visual task, so the designer might have to illuminate them using a series of recessed or hidden, more technical luminaires. It is not necessary that they help in the actual lighting of the interior, but if they do, the designer can make use of them and build the system around them.

Visual comfort is maintained by verifying the luminance output of the fitting. This can be done by looking at luminance diagrams, and by having the simulation software calculate luminance contrasts to verify that they do not go beyond the required values.

26.2 Task layer

Solving the decorative layer means creating constraints for the other two levels: there may be a certain level of background luminance to maintain, or newly identified visual tasks to solve.

To solve this layer and satisfy these constraints, the designer will select those light fittings that satisfy the standards' requirements for each visual task.

Layered design

One way is to adopt a specific type of luminaire for each task, e.g. selecting a wall-washer to illuminate a bookcase, a downlight for a table, and so on. Another is to make use of fittings that solve other layers, for example a decorative torchères might illuminate with indirect light an easy chair where one might want to read a book; or a fitting that provides direct light to an office might also illuminate the desk.

Activity	Location	Requirements			
		lx	CRI	U_0	UGR
Food preparation	Kitchen	500	80	0,6	22
Visual tasks	luminaire	compliance check			
Counter top 1	Philips eW Powercore model BCX413 10W	•	•	•	•
...	...	...	...	...	...

Table 34: Activity table updated with luminaires and compliance checks.

Once all the tasks have been solved, the interior will be lit where each task is, and dark elsewhere.

This is where the Activity Table is updated by including the luminaires used to illuminate each activity (see Table 34). The description of the luminaire is very important: make, model and power must be included and compliance should be checked.

26.3 Brief note on regulations and standards

We know that the main European standard related to interior is the EN12464-1:2011. They deal with the illumination of work places, and give back a set of minimum requirements for each activity that is performed in the interior. In particular, the minimum maintained illuminance, the minimum colour rendering index, the minimum uniformity and the maximum glare rating. The idea is that the more complex the activity, the higher the minimum requirements.

If the activity in question is not directly work-related, it is wise to compare it to the equivalent professional activities and assume the same requirements, when visual performance and comfort are important. So, for example, a home kitchen should have the same requirements as a restaurant one, a TV room or a home office the same requirements (including luminance control) of an office with DSE, and so on.

The absolute minimum requirement in uniformity is $u_0 > 0.10$.

The minimum maintained illuminance in area where one is expected to remain a long time is 100 lux; for a place where the stay is temporary the requirement is reduced to 50 lux.

26.4 General layer

In order to solve this layer and complete the lighting system in the interior, the designer will have to integrate the light given to the tasks by adding light fittings. The idea is to provide the light that allows people to use and move about in the interior, no matter where the activity is.

There are places conspicuous for their apparent lack of general lighting: clubs where almost everything is in the dark and the lighting is mostly used for effects, or restaurants where each table is a pool of light, maybe candlelit, in a dark room. The point to consider here is that these interiors have multiple operating modes: for the restaurant "atmospheric dinner" is one, but there is also "cleaning and maintenance", or "work lunch", and in these other modes general lighting is a must. When the cleaning and maintenance staff is at work, or during a work lunch, neither the customers nor the staff will appreciate the ambiance lighting that dinner customers will love.

26.5 Self evaluation

Answer the following questions to test your understanding of the material:

Layered design

1. Describe a possible work flow that would bring the designer to become familiar with the interior and its different aspects and requirements.

2. Describe the Complete Activity Table, its function and how it is compiled by the designer.

3. Describe the Layered design method, detailing properties, challenges and decisions to take for each layer.

4. Select a luminaire at random from the Internet, invent a task that would fit, then see on what layer it should go. Repeat a few times

Part 5

This part of the book deals with the intelligent use of a lighting calculation programme.

I selected DIALux, available free of charge from DIAL (www.dial.de). It is unfortunately a Windows only programme, but it works with the most common virtualisation programs, such as Parallels or vmWare, on the Mac.

This book refers to version 4.12. I chose not to use the latest (EVO) version, even if the interface is much more modern, for two main reasons. First because at the time of writing it does not yet do everything that version 4.12 does; second because DIALux EVO requires an OpenGL 3 compatible graphics card, which some readers might not have, and is not supported by virtualisation packages for the Mac.

This part of the book does not aim to replace the manual available for the program: I will not spend a lot of time describing how to do something (this is where the programme's manual comes in), but I will try and explain why that something is done, what it is for and when it is useful.

Introduction to a lighting calculation program: DIALux

Modern light calculation programmes are really powerful, relatively easy to master, and they give precise quantitative results. This means that it is now easy to verify compliance and visual comfort with a realistic simulation of the interior, and evaluate the concept from a functional point of view.

27.1 The objective

> The **reason** why we use lighting calculation programs is to design **better** lighting systems. (63)

This may sound obvious, but it highlights how the software must be practical, effective and based on the real physical behaviour of light.

This is the main difference between a lighting calculation software and a photo-realistic rendering software: the former outputs a realistic result, the latter needs to produce a realistic **looking** result.

A lighting calculation software can help design a system in two ways.

1. It can give the designer quantitative results, so if one is designing an office and one does not get at least 300 lx of average maintained illuminance on the programme, there is then something wrong with the project.
2. It lets the designer see the result of changes in the system quickly enough that multiple solutions can be analysed.

To go back to the previous example, let's say one got 250 lx in the office. What is the best way to get to 300 lx? Should one increase the number of luminaires, or select a more powerful type, or reduce the mounting height? If the lighting calculation software can quickly produce results, the designer can test each of these solutions and find the best one.

27.2 The programme

A lighting calculation programme is made of four different parts working together:

1. A 3D modeller.
2. A database of technical data, descriptions and images of luminaires
3. A simulator that does the actual calculations.
4. The package that composes and produces the output

27.2.1 3D modeller

The 3D modeller lets the designer create a realistic model of the interior and its exterior. The idea is that it can replicate:

1. The interior: including
    1.1. Windows and doors
    1.2. Furniture
    1.3. Textures and colours
    1.4. Various architectural features
2. A description of what is outside.

Modelling with an interactive simulation software (such as any lighting calculation programme) faces the designer with the problem of finding an equilibrium between precision and simplicity.

Precision is a requirement because if the model is not faithful to the real interior, the results will be irrelevant.

Simplicity is a requirement because of the interactive nature of the process.

Let;s start by looking into simplicity.

The model built in DIALux will look very different (just a few large surfaces) compared to one that would work with a 3D rendering software. For example, a desk with rounded corners may be approximated to a rectangular one without loss of important details in the lighting software, while it would of course be unacceptable in a photo-realistic rendering software.

The reason for this comes from 27.1 on page 317: the software calculates quantitative data for all the surfaces of the interior, so the fewer the surfaces the quicker the calculation. Considering that modelling a lighting system is an iterative process, it is important that the simulation time is kept to a minimum: if the designer has to wait for ten minutes to get the result of a small change in the interior, he will make fewer changes. On the other hand, if the calculation time is reduced to 10 seconds, the designer can make more modifications in the same amount of time, and since for every few changes on average there is one improvement, he will design a better lighting system.

This is completely different from using a simulation program that is not interactive (such as a photo-realistic rendering software): it allows the designer to for example start the calculation in the evening and get the result the morning after without any problem.

There is a second reason as well. Since we are mostly interested in average values on flat surfaces (usually the average maintained illuminance) for example on a desk, and since the program calculates these values for each surface, it makes sense for the top of that desk to be made by a single unbroken face, rather than multiple small ones, so that one can immediately obtain the numerical results for it.

At this point it should be apparent why one shouldn't just import the model built for the photo-realistic rendering software into the light calculation software.

In order to be more effective, the designer should use as many items in DIALux's furniture database as possible, scaling them so that the dimensions are exactly like the ones of the real objects.

Alternatively, and this is my advice, one should use as many cubes and boxes as possible, correctly sized, instead of objects in the interior, because they have very few polygons, so the calculation is fast; all their surfaces are flat, so there is no need to manually add more on which to display results; and their approximations of real objects is often quite acceptable.

> ```
> The objective of the model is to find a
> compromise between precision (include all
> the information that is needed to obtain
> a realistic effect for lighting) and sim-
> plicity (no unneeded details).
> ``` (64)

To continue the comparison to the rendering software, let's look back to the work-flow we follow:

The rendering software lets the designer visualise his concept in a photo-realistic way for himself and for the client using impossible solutions (e.g. lights that suddenly cut off at a certain distance from the source, ambient lighting independent of light sources, and so on). It is of enormous importance not only to communicate the result, but also to help the designer see his choices.

Once the concept is perfectly clear, the next step is to find a way to implement it in reality. The designer must find the correct luminaires, positions and orientations that will give him the result that he has in mind.

Introduction to a lighting calculation program: DIALux

At the same time, the lighting system must function, in the same way that a curtain, a wall, a window must function and not damage or hurt the user.

This is where the requirement of precision is relevant. Dimensions, reflectances and modes of reflection, colours, the types and properties of the luminaires, any transparency, orientation of the building, position of windows and doors etc., must be faithfully represented in the programme. A few centimetres of difference in the position of a light source may not mean a lot for a photo-realistic rendering software, but it will make a significant difference for a lighting calculation one.

27.2.2 Database

DIALux contains three databases, one for objects, one for colours and textures and one for luminaires. The first two are found on any 3D modelling program, the third one is the more relevant to lighting.

The object database contains primitives (cubes, wedges, and so on), elements of the room (windows, doors, columns, etc.), elements for exterior scenes and obstructions (trees, cars, buildings), and furniture.

The textures and colour database contains materials for indoor and outdoor objects sorted by type (concrete, wood, etc.) or by application (floor, ceiling, etc.), a set of colour coded with the RAL system, a set of colours for lights sorted by colour temperature, and colour filters for light sources.

The luminaires database is sorted by manufacturer, and contains dimensions, pictures and technical data (photometric solid, types of lamps, flux emitted) for each luminaire.

There are three ways to put a luminaire into the programme.

The easiest way is to check the manufacturer's website to find out if they have a plug-in available for DIALux. The advantage of plug-ins is that they usually have pictures, descriptions of the fitting, and search engines that help select a luminaire. About two hundred manufacturers have a plug-in.

Failing that, some online catalogues of manufacturers have a button to automatically insert the luminaire's technical data into DIALux.

Finally, the designer may have to resort to downloading the actual luminaire files from the manufacturer and import them into the program. Luminaire Data Files (LDT) are usually in either Eulumdat format (.ldt) or IES format (.ies). Both these files contain the dimensions of the luminaire, the power and luminous flux emitted, the lamps installed, and the photometric solid of the luminaire.

27.2.3 Simulator

The simulator is the part of the programme that actually makes calculations. DIALux can calculate indoor and outdoor scenes, road lighting, and most importantly it can include daylight in the calculations.

Daylighting (see "32 Daylighting" on page 355) is really important in today's lighting design. It is a free source of very high quality very powerful light, so the challenge is to make as much use of it as possible to reduce costs, without damaging visual comfort.

27.2.4 Output package

The output package is able to produce a very large amount of data: for a small project it is easy (and unwise) to generate over a thousand pages. It is obviously quite important that the designer operates a careful selection of what information he should produce.

Normally the designer should deliver a project output that contains enough data to fully describe the interior, the lighting system and the results. It is also wise to include a note containing details on the choices made (see Table 35 on page 323)

The output must describe the interior: it should contain the size, the position, the shape, the colour material and finishing data for each room, window, door, furniture and object inside. The objective here is to document how the interior simulated is equal to the real life one.

Table 35: Detailed output.

For each:	Output
Notes	Details of the designer's choices
Luminaire	Technical data sheet
	UGR table (if available)
Room	Floor plan with luminaires
	Luminaires layout plan
	Position and sizes of calculation surfaces (any type)
	Control groups (if present)
Control group	Planning data
Light scene	Planning data
	Summary
	Photometric results
	Calculation surfaces
Control surface	Isolux, minimum, maximum and average maintained illuminance

The output must describe the lighting system as well: all the properties of the luminaires chosen, and of the system. So for each luminaire the output must include the type, the lamp employed, the position and orientation, the light intensity distribution, UGR, and, if needed, the luminance table. For the whole system the output must contain details on all scenes and control groups (see "31 Control groups and light scenes" on page 347), layout plans, data on position and size of all the calculation surfaces (see "30 Calculation surfaces" on page 341). The objective here is to describe a system not only to the client, but also to the professional who will actually install it.

Finally, the output must contain data on illuminance (usually isolux curves, average maintained, maximum and minimum), luminance, uniformity and UGR to describe both the compliance of the system to the regulations, and the visual comfort obtained.

27.3 Self evaluation

Answer the following questions to test your understanding of the material:

1. Describe how a lighting calculation software can be useful to the designer, and compare its objective to those of a photo-realistic render programme.
2. Describe the components of DIALux.
3. Explain how one can import luminaires into DIALux.

Introduction to a lighting calculation program: DIALux

DIALux Wizards

28

Wizards are tools that make it easier to complete simple projects, where it is important to get a quick result, or when the interior is standard and requires little customisation.

28.1 Use of a wizard

The advantage is that the designer needs very little experience to complete a lighting design project quickly and efficiently. The project can then be edited to add personalization and create a complex result.

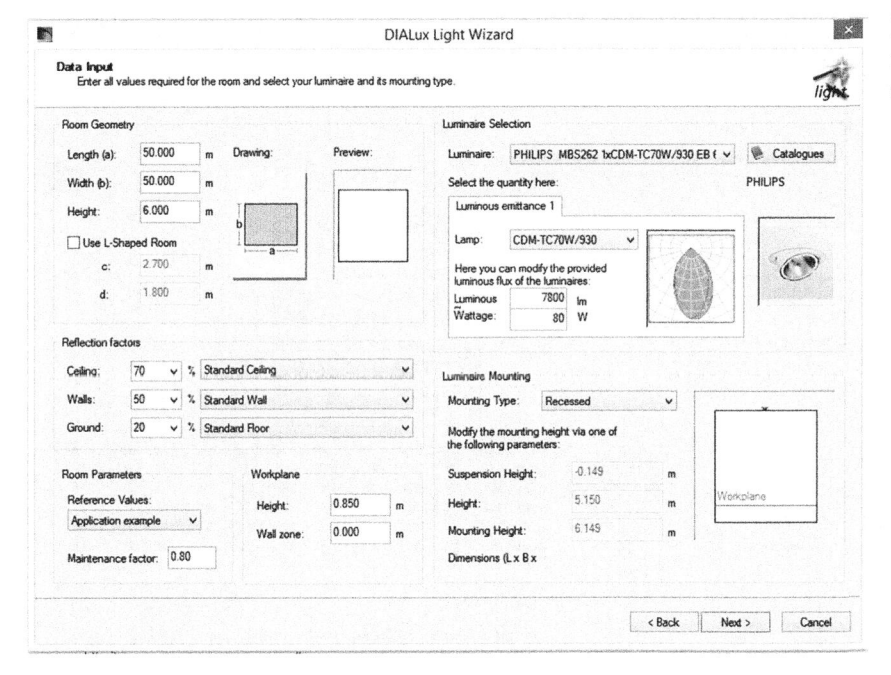

Figure 217: Data mask for DIALux light wizard.

One of the most interesting wizards is DIALux light. The following paragraph contains an example developed with it.

Figure 218: Calculations mask for the supermarket on DIALux Light.

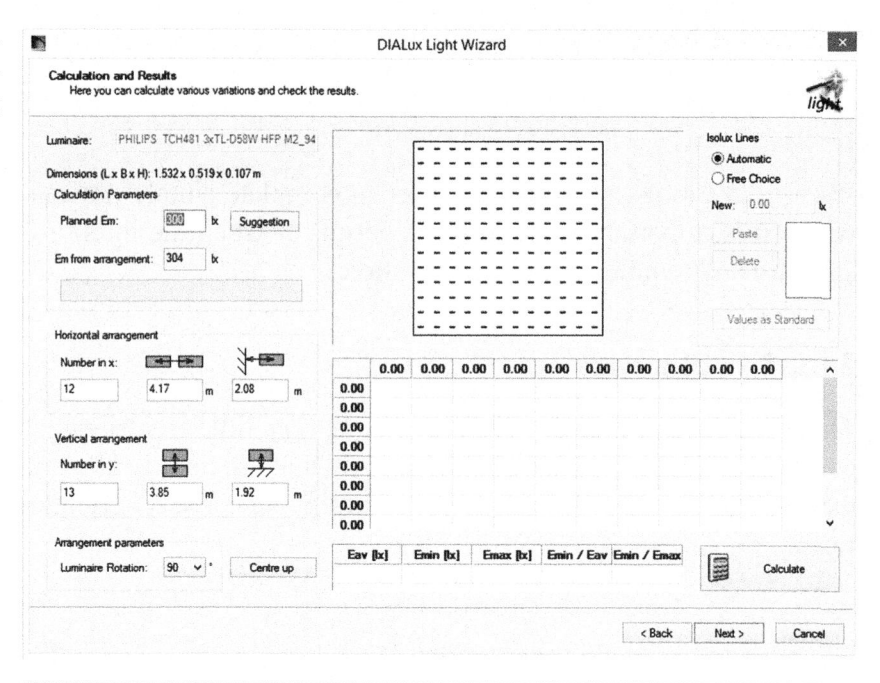

Figure 219: Calculation mask showing the results.

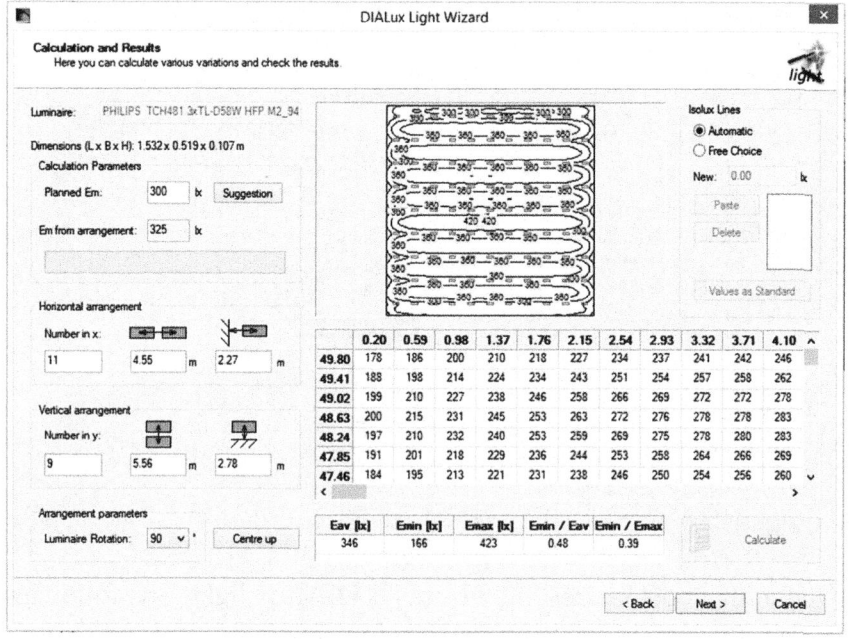

28.2 An example: the supermarket

In this quick project I wanted to find out how many 70W metal halide recessed downlights are needed to produce 300lx in a supermarket. It is a square room with a side fifty metres long and six metres high.

Figure 217 on page 327 shows how, in a single mask, the wizard shows all relevant data for the project.

From the top left going down we see the size and shape of the room, the reflectances for ceiling walls and floor, the maintenance factor and work-plane height. In the right half of the mask there are details on the luminaire selected and on its mounting.

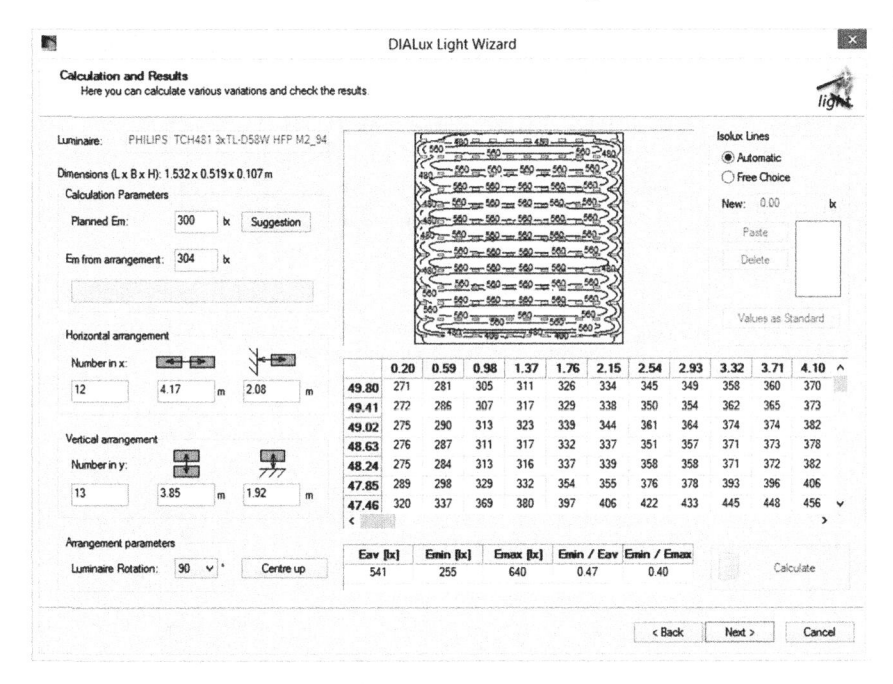

Figure 220: Updated results for the supermarket..

Figure 218 shows the calculation mask for the project. On the top left the designer can input the average maintained illuminance he would like to obtain, and the program suggests an arrangement of luminaires, detailed on the bottom left part of the mask. On the top right of the mask there is a plan view of the interior and an illuminance table.

I selected 300 lx, pressed **calculate**, and got the result in Figure 219. The right side of the image now shows isolines, the table is populated and the last row shows the average maintained, minimum and maximum illuminance, with two types of uniformity.

It is now really easy for the designer to change any input and have the program recalculate the results. For example, I changed the required average illuminance to 500 lx, and obtained the results in Figure 220.

The last mask for this wizard is the output mask (see Figure 221), where the designer can quickly decide what to present and what to do with the project.

Figure 221: Output mast for the DIALux Light Wizard.

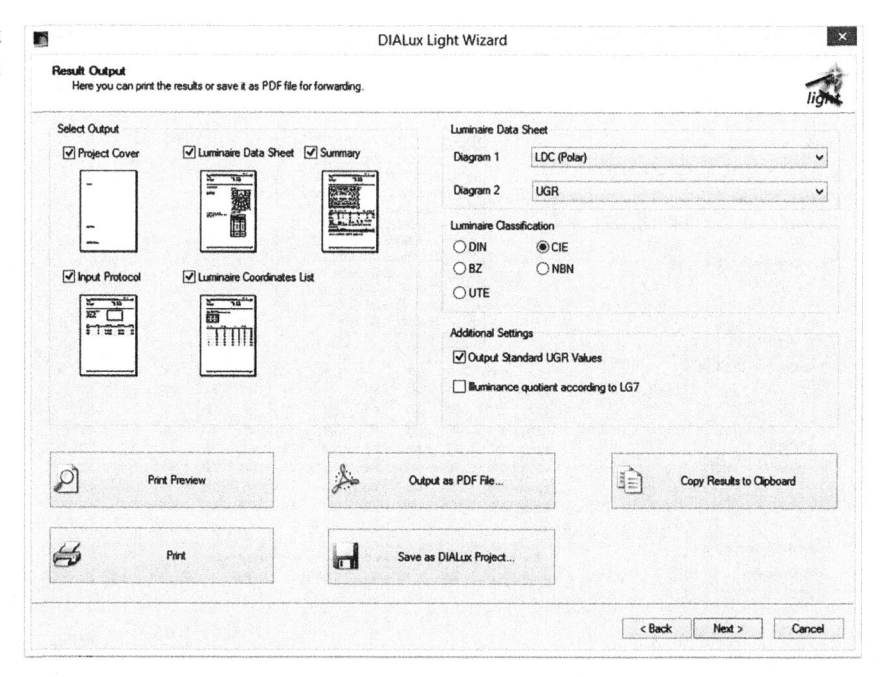

28.3 Self evaluation

1. What are wizards?
2. Invent a project in which a wizard can be used effectively, and another in which a wizard would not work

Figure 222: Side view of the classroom used as sample interior.

Figure 223: Plan view of the classroom used as sample interior.

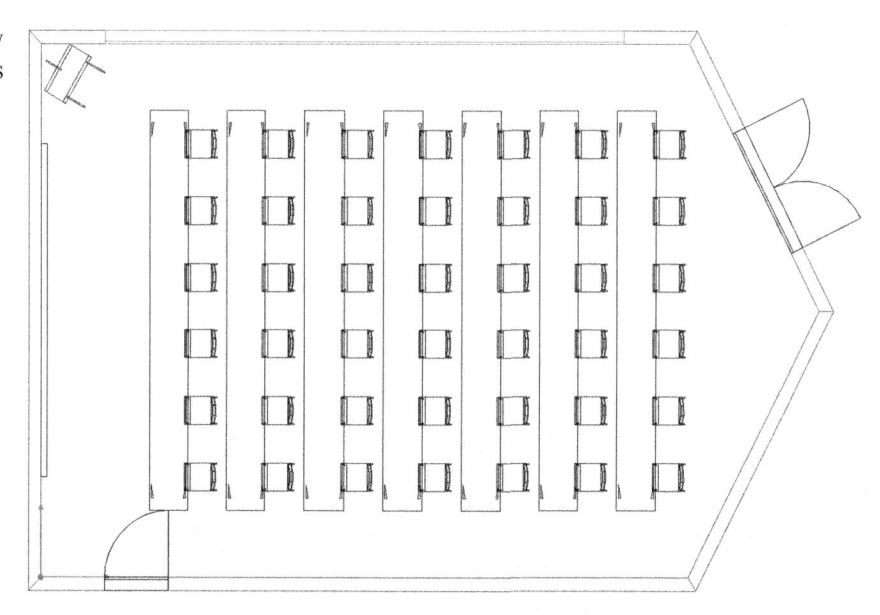

Figure 224: View of the obstructions of the classroom.

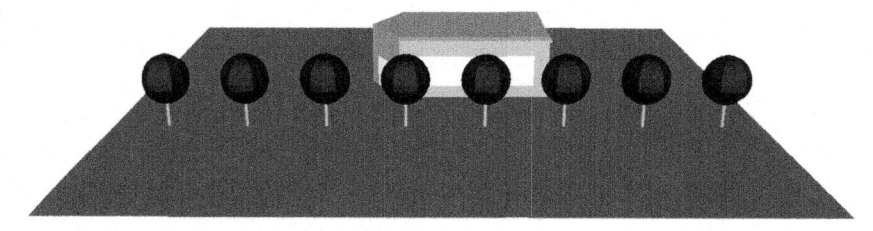

A sample interior

29

Let's follow the indications in this chapter to build a model of a sample interior. It's better to have the programme open when following the tutorial starting with 29.3 on page 334.

29.1 Let's build the interior

The room is pentagonal, 12 metres by 8, with a large window on the north wall and two doors. The result is in Figure 222 and Figure 223.

Because this interior has large windows, we need to model the obstructions as well. The result is in Figure 224 on page 332.

The interior is on the ground floor, it has a row of trees in front of its large window, and a small overhang.

A very important thing to remember is to rename every object added into the project with a significant name, so that the designer can recall its position and properties by name only. The importance of this step is often missed, until the designer must select the correct luminaire among a list of **Line01**, **Line02**, **Line03** and so on, or the correct desk in a list of fifty-seven **Cube01** objects.

29.2 Objective

Draw the model of a classroom or conference room interior, on the ground floor of a building in a square 40m by 40m, with an overhang and a row of trees in front of the window.

Refer to the manual[1] of the programme as needed. It is available from dial.de:

---

1. Specifically the parts dealing with the user interface, creating a new project, editing a room, and inserting room elements, furniture and textures.

After you type a value in DIALux, always press the **enter** key, or it may not be updated

29.3 Create the room geometry

`Start -> All Programs -> Dialux -> Dialux 4.12`
Choose `New Interior project`
`Length`: type 10, `Width` type 8, `Height` type 4
Click `OK`
Name: Replace `Room1` with `Classroom`
Click on `Floor plan` icon
Right click on `Classroom`
Click on `Edit Room Geometry`
Click on row number 3, column x
Click on `Insert coordinates`
Type 12 on `x` and 4 on `y`
Click on `OK`
Click on `Overall View of the Scene` Icon

29.4 Add doors and window

Select `Objects` tree
Click on `Windows and doors`
Click and Drag `Window` on North wall (`Wall 4` in `Project` tree)
Click on `Position/Size`
`Width`: type 8, `Height`: type 2, `Distance from Left`: type 1

Click and Drag `Door` onto south wall (`Wall 1` in `Project` tree)
Click on `Position/Size`
`Distance from Left`: type 8

Click and Drag `Door` onto North Eastern wall (`Wall 3` in `Project` tree)
`Width`: type 2, `Distance from Left`: type 1.5

A sample interior

Click on **General**, on `type of opening` select **Outward, double wing**

29.5 Textures

Click on `3D Standard View`
Select **Colours** tree
Click on **Textures, Indoor, Floor, Wood, Strip flooring**
Click and drag `Oak wood-12` onto the floor

Click on **Textures, Indoor, Wall, Plaster**
Click and drag `Roughcast plastering white` onto the wall

Click on **Textures, Indoor**, Ceiling
Click and drag `Ceiling panels` onto the ceiling
(if you can't see the ceiling, click on `Rotate View`, then click and drag on the 3D View window to rotate the view)

Click on **Textures, Indoor, Doors**
Click and drag `Door, wood maple` onto both doors

Click on **Textures, Indoor, Window**
Click and drag `Window, Wood` onto the window

Select **Project** tree
Click on the window in the 3D View
Click right on the tab twice, to get to the **Texture** tab
Select the **Texture** tab
`Size Y` type 2

29.6 Insert furniture

Click on `Floor plan` icon
Select **Objects** tree
Click on **Object files, Indoor, Furniture, Tables**
Click and drag `100x60 standard` onto the floor

Position of the object origin: X type 2, **Y** type 4
Size: L type 6
Rotation: Z type 90

Click on **Object files**, **Indoor**, **Furniture**, **Chairs**
Click and drag **simple chair** onto the floor
Position of the object origin: X type 2.5, **Y** type 6.5
Rotation: Z type -90

Click on **3D View** icon
Select **Colours** tree
Click on **Textures**, **Indoor**, **Furniture**
Click and drag **Ash light grey** on both the table and the chair

Select **Project** tree
Select **simple chair**
Right click and choose **Copy along a line**
Number of copies type 5
Distance: X type 0, **Y** type -1
Click on **Copy**

Shift-select everything under **Objects** (**100x60 standard**, and 6 **simple chair**)
Right click and choose **Copy along a line**
Number of copies type 6
Distance: X type 1.2
Click on **Copy**

Select **Objects tree**
Click on **Object files**, **Indoor**, **Office accessories**
Click and drag **Flip Chart** onto the floor
Select **Geometry** tab
Position of the object origin: X type .5, **Y** type 7.5
Rotation Z type 60

Click on **Standard elements**

A sample interior

Click and drag **Cube** into the room
`Position of the object origin:` **X** type 0.05, **Y** type 4, **Z** type 1.3
`Size:` **L** type 0.1, **B** type 5
Select **Name** tab
Replace **Cube** with Whiteboard
Select **Colours** tree
Click on `Colours, 9xxx Black/White`
Click and drag `9016(Traffic white)` on the `whiteboard`

29.7 Surroundings of the building

Select **Project** tree
Right click on `classroom`
Click `Edit daylight obstruction`
Click `Obstruction of Project 1`, Ground elements
Right click on `Ground Element 1`
Select `Edit Ground Element`
`Length` type 40, `Width` type 40
Click on **OK**

Select **Objects** tree
Click on `Object files, Outdoor, Trees`
Click and drag `Tree01` onto the ground
Select **Geometry**
`Position of the object origin:` **X** type 24, **Y** type 14
Select **Project** tree
Right click on `Tree01`
Select `Copy along a line`
`Number of copies:` type 7
`Position of the last copy:` **X** type -10
Click on **Copy**

Select **Objects** tree
Click on `Standard elements`
Click and drag **Cube** onto `Obstruction of Classroom`
`Position of the object origin:` **X** type 5, **Y** type 8.5,

Z type 3.95
Size: L type 10, **H** type 0.1
Select **Name** tab
Replace **Cube** with Overhang
Select **Colours** tree
Click **Textures, Outdoor, Wall, Concrete**
Click and drag **Slabs light concrete-25** onto **Overhang**

Select **Project** tree
Click on **Project 1**
Select **Location** tab
Location: Select **Milan**
Click **Save location**
Click on the **File** menu
Save As... choose your file name in a meaningful way, i.e. your surname and the title of the project.

A sample interior

Figure 225: Side view of the classroom used as sample interior.

Figure 226: Plan view of the classroom used as sample interior.

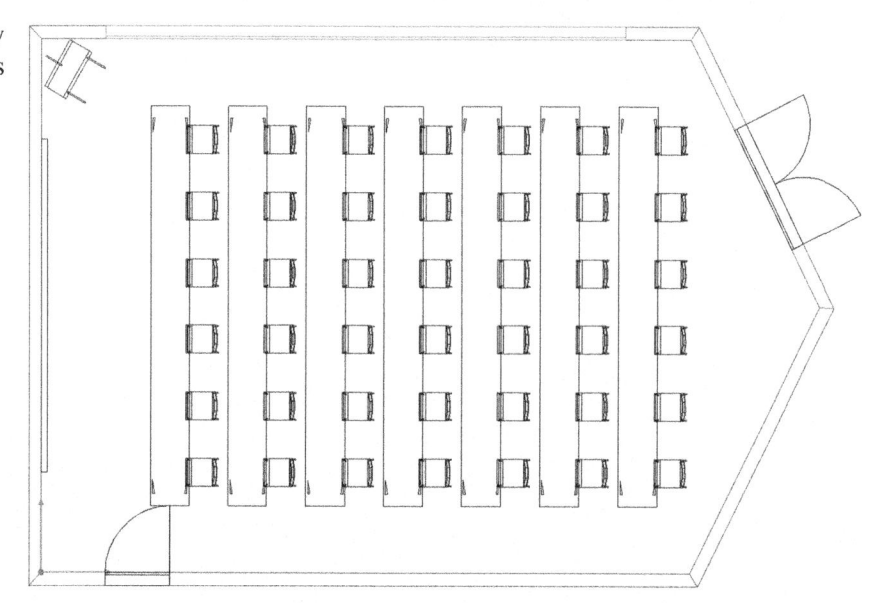

Figure 227: View of the obstructions of the classroom.

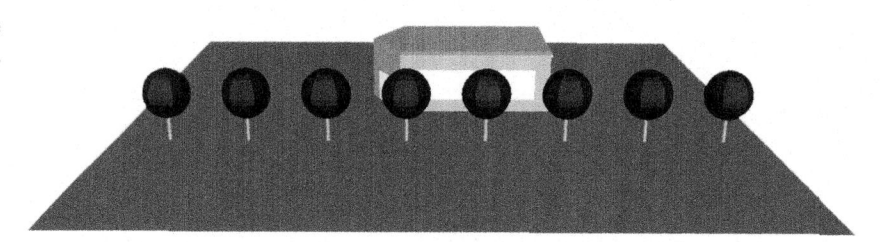

Calculation surfaces 30

We are interested in seeing the results of the lighting system on flat surfaces: desks, work surfaces, walls, ceilings and wherever a task is. So we need to tell the program that we want to see these results on the surfaces we choose.

30.1 Walls, floor and ceiling

The results on walls, floor and ceiling are automatically calculated by DIALux, and they can be presented in the output. The real question is whether they are relevant to the project:

1. Walls and ceilings are usually only significant when they host a task that is about as large as they are;
2. The floor is relevant only in those areas where there is no furniture. Isolux curves still have meaning on a floor partially covered with furniture in those spots that are clear of items.
3. Averages, maximum, minimum and uniformity values become almost irrelevant because they are calculated including the shaded areas under the furniture. This reduces the average and the minimum values, and can change where the maximums are. Consequently the values for uniformity will be worse than they really are.

30.2 Work plane

In order to reduce the effect of furniture, the program calculates values on a non physical surface called work plane. The designer can modify its height and its distance from the walls.

Another reason to use the work-plane is when the actual task areas aren't defined yet. For example, in the supermarket of 28.2 on page 329 we might want to know what is the average illuminance on the work plane even before deciding where to put the counters and the shelves.

30.3 Surface of an object

The program calculates data for all the surfaces of any object in the room, so it is only a matter of telling DIALux to output these results.

Figure 228: Dashed line highlighting the border of the selected surface. On the left, the front of the flip chart is selected; on the right, the left side of the flip chart is selected.

The way to do it is to right click on the surface of the object where one needs the data displayed, choose **Select this surface**, then, on top of the **Project** tree select the **Calculating Grid** tab and put a check on **Output results**.

It is important to make sure that the surface selected is the correct one. DIALux highlights the object selected in red, and the surface with a dashed line (see Figure 228). The reader should do the same with the front of the whiteboard and the surface of the first desk.

30.4 Free surfaces

We may need to find out what results we obtain on a surface that is not part of an object and that does not affect the illuminance distribution under it.

The typical example is when we are illuminating an interior where the positions of the task areas are not certain. Let's say we design the lighting system for the temporary exhibition area of a museum, we don't know where the paintings will be placed, or how large they might be, so we could place calculation surfaces on the walls (**Object** tree, **Calculation surfaces**, **Calculation surface**), to show the effect of the spotlights we use.

30.5 Task area and surrounding area

A lighting system that is compliant to regulations and produces visual comfort will not only illuminate the task, but the surroundings as well. To make it easier to display these results, DIALux has an object called **Task Area** which is made of a task area and a surrounding area.

The designer just needs to place it on the task and dimension it.

Let's drag a task area into the sample project, and look at the **project** tree, where it is shown to be made of two objects: one called **Task Area 1** and another called **Surrounding Area**. For **Task Area 1** we will set the **Position of the object origin X**: at 8 metres and **Y** at 4 metres, while the **Size** will be 0.6 m by 6 m. For **Surrounding Area** we will set the **Position of the object origin X**: at 7 metres and **Y** 4 at metres, while the **Size** will be 4 m by 7 m. The result is in Figure 229.

Figure 229: Task area and surrounding area.

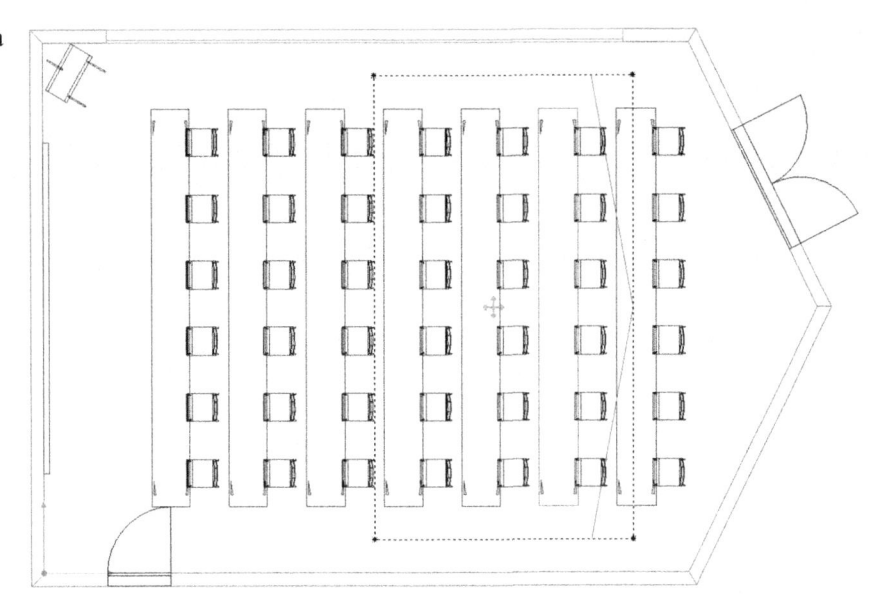

30.6 Self evaluation

Answer the following questions to test your understanding of the material:

1. Explain what is a calculation surface.
2. Describe each subtype of calculation surface.
3. Invent an example that requires each subtype of calculation surface.

Figure 230: Side view of the classroom used as sample interior.

Figure 231: Plan view of the classroom used as sample interior.

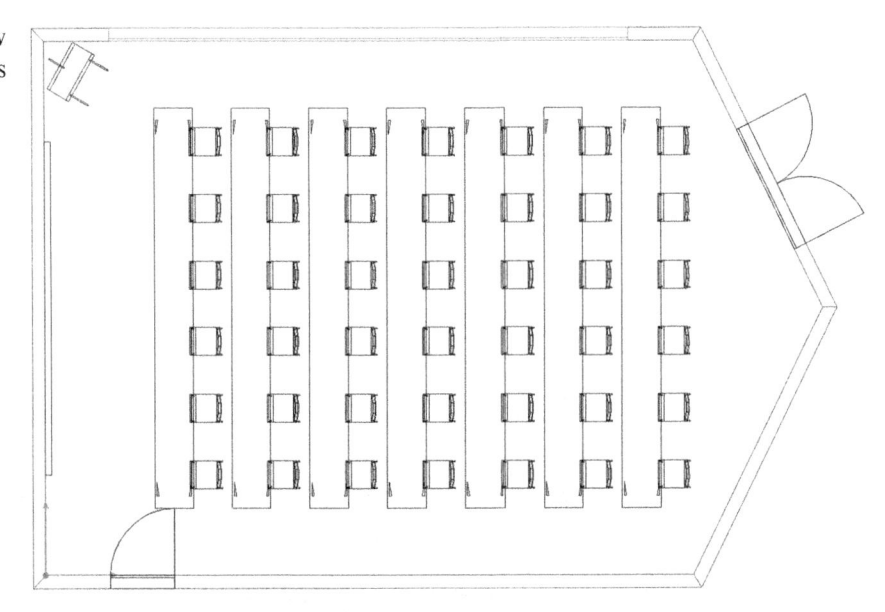

Figure 232: View of the obstructions of the classroom.

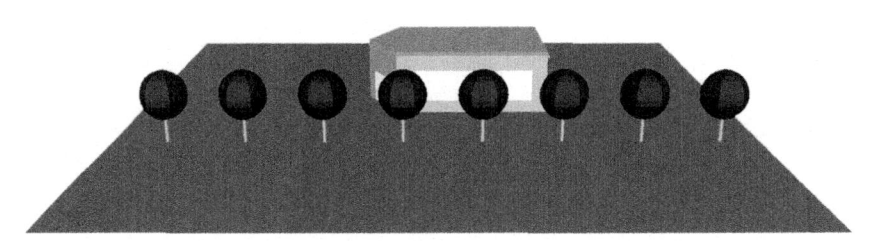

Control groups and light scenes

31

One of the most important steps in designing the lighting system is the familiarisation with the interior, in particular with its different uses. They often change with time, and may have different requirements for the lighting system.

31.1 What is a light scene?

Let's think about illuminating a restaurant. After a detailed interview with the restaurant manager, the designer deduces certain informations:

1. The restaurant is large and split into multiple rooms,
2. It is open for lunch and dinner.
3. All the rooms of the restaurant are in use at dinner, while during lunch only a few of them are open.
4. The lunch crowd is mostly composed of office workers. The dinner customers enjoy a dimly lit atmosphere.

The designer identifies three modes of operation for the interior. Before reading further, please try and identify them yourself.

They are:

1. Lunch.
2. Dinner.
3. Cleaning and maintenance.

The lighting must reflect the different usese of the interior: the night customers might enjoy their tables dimly lit and the rest of the interior as dark as possible, the staff needs high illuminance everywhere, and the lunch crowd will probably prefer a more natural light.

The lighting system must satisfy these different uses of the interior, so we need to create three scenes, one for each use.

> ```
> A light scene is the setting of the light-
> ing system that corresponds to a real work-     (65)
> ing condition of the installation.
> ```

The designer must decide the luminaire arrangement with the scenes already clearly defined.

The work flow will be:

1. Identify the different situations (for the restaurant: "cleaning and maintenance", "lunch" and "dinner").
2. Create a scene for each work mode of the installation.
3. Group the luminaires accordingly.

31.2 What is a control group?

Even in a small lighting system it is unreasonable to expect that each luminaire will have its own switch. Very often multiple luminaires are joined under the same control because they share the same function, or are connected logically: there never is a situation in which a few of the luminaires in the group are on and the others off.

> ```
> A control group is a series of logically
> connected luminaires that share the same        (66)
> control.
> ```

The system behaves as if the luminaires part of a group were al connected to the same switch/dimmer, even if this does not necessarily mean that the actual real life luminaires are part of a single circuit.

In the example of the restaurant, we might have the luminaires providing general lighting as part of a control group for each room, such as "general lighting 1", "general lighting 2" and so on. The same might happen to table highlighting luminaires: each room's

Control groups and light scenes

light fittings may be part of the same control group: "table highlighting 1", "table highlighting 2" and so on.

The different controls can be joined into a control panel. Figure 233 shows the one for the restaurant.

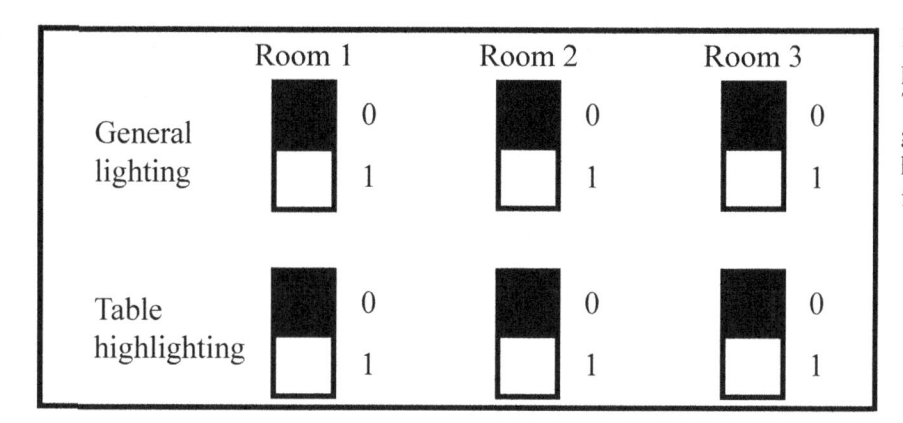

Figure 233: Control panel for the restaurant. There are six control groups, three for table highlighting and three for general lighting.

31.3 An example: lighting a restaurant

In the previous two paragraphs we identified three scenes for the restaurant:

1. Lunch scene: It contains no special lighting or table highlighting, only the first two rooms of the restaurant are in use.
2. Dinner scene: All the rooms are used, table highlighting only.
3. Cleaning and maintenance: all the areas fully lit.

The control panels for the lunch scene is shown in Figure 234. General lighting is on for the first two rooms, and table highlighting off for the whole restaurant.

Figure 235 shows the panel for the dinner scene: General lighting is off and Table highlighting on everywhere.

Figure 234: Control panel for the Lunch scene.

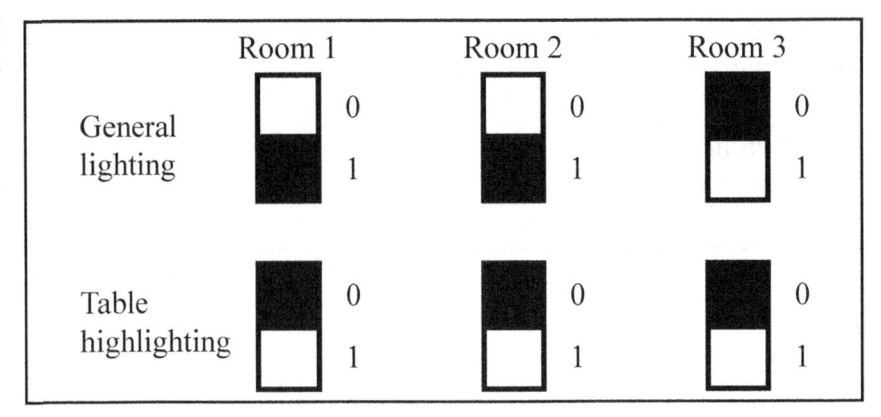

Figure 235: Control panel for the Dinner scene.

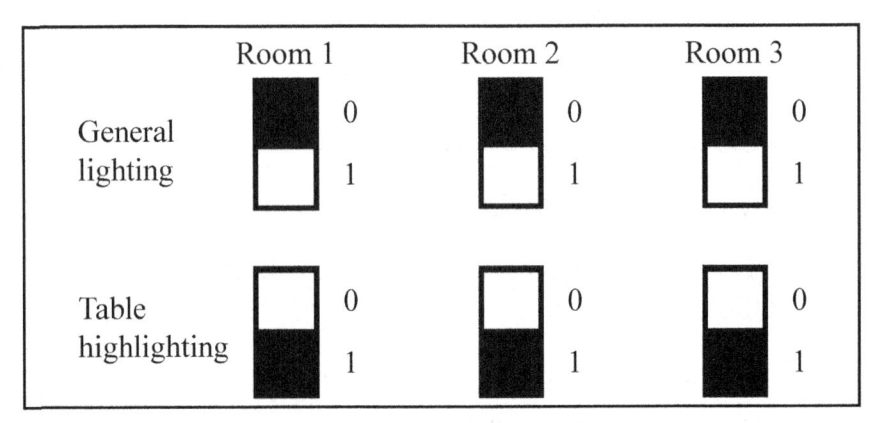

Figure 236: Control panel for the Cleaning and maintenance scene.

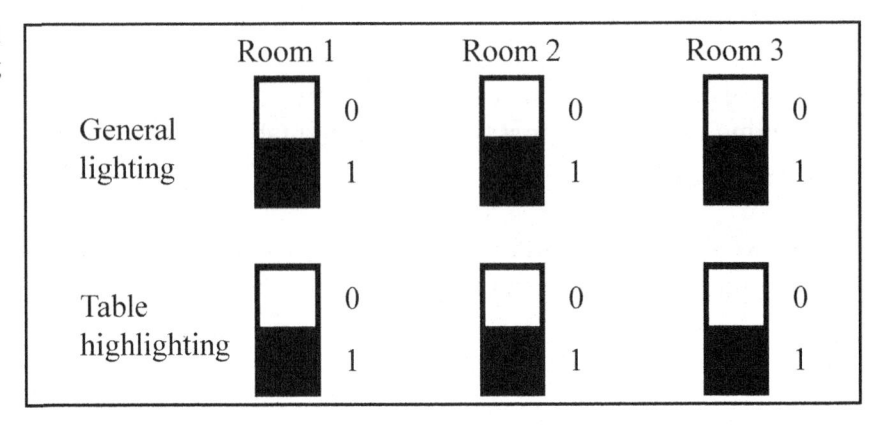

Control groups and light scenes

Finally the control panel for the cleaning and maintenance scene is in Figure 236. Everything is on, because the staff needs full lighting to complete their task.

DIALux does not have a control panel like the one in Figure 233, it uses a slightly different graphical layout, as shown in Figure 237. On the bottom halves of both pictures there is the light scene's name (respectively Lunch and Dinner) with the control groups connected to it.

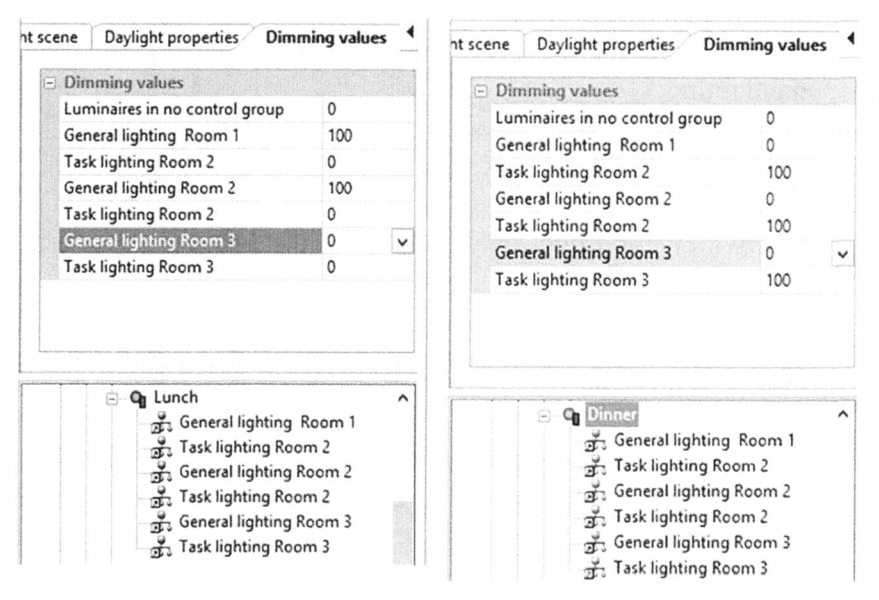

Figure 237: Lunch (left) and dinner (right) light scenes for the restaurant in DIALux.

On the top half of the images there is the detail of the light scene, showing the dimming values for each control group.

DIALux has a default control group, called `Luminaires in no control group` that is sometimes useful to simplify the splitting of the system under different controls. Finally the dimmer for each control groups allows a percentage of the full flux to be emitted, rather than being a simple on-off switch.

As we know from Part 2 on page 121, not all light sources can be dimmed, so the designer must make sure he is selecting the right luminaire.

31.4 Application to our sample interior

Looking at our sample interior in Figure 230 on page 346, we should notice the large window on the North wall. This tells the designer that natural light can play a role in illuminating the interior, and, if that is true, then the lighting system must be split into two control groups:

1. One with the luminaires that need to be on, no matter how much daylight there is.
2. Another with the luminaires that can be turned off to exploit daylighting.

This of course mean that the system should have at least three scenes:

1. One with natural light only, to find out what part of the interior can be naturally lit,
2. One with both natural and artificial light,
3. One with artificial light only.

All this seems common sense, but the designer needs to make sure, so he needs to test these assumptions with a simulation to calculate the daylight scene and see how well natural light illuminates the installation.

31.5 Self evaluation

Answer the following questions to test your understanding of the material:

1. What is a control group?
2. What is a light scene?
3. Invent an example of a lighting system requiring at least three light scene and describe it in detail.

Figure 238: Side view of the classroom used as sample interior.

Figure 239: Plan view of the classroom used as sample interior.

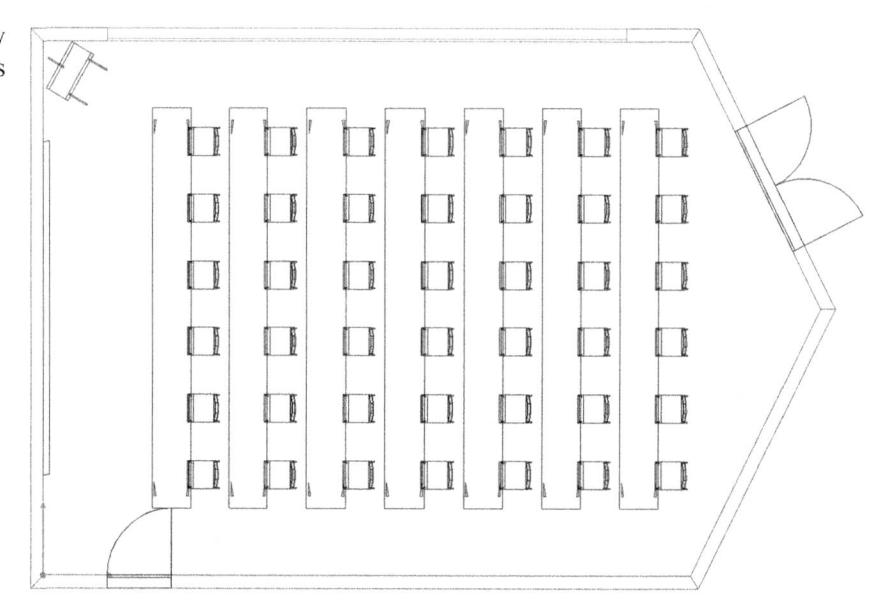

Figure 240: View of the obstructions of the classroom.

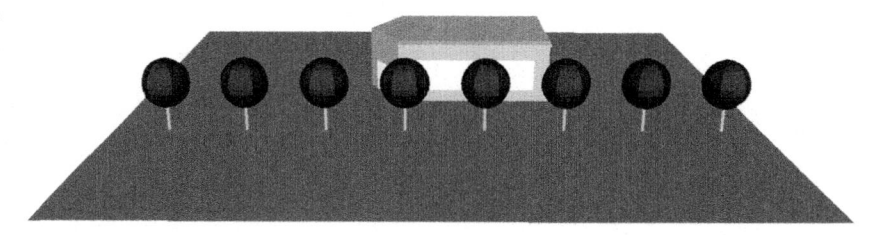

Daylighting

Natural light is abundant, up to 120000 lx; it has varying colour temperature, from 2000 K at dawn or sunset to 25000 K of a clear sky with a cloud in front of the sun; and it has fantastic colour rendering, after all our eyes evolved to see colours under that same light.

In short, it is very desirable. The issues it has come from poor visual comfort: the luminance, illuminance and luminous flux of natural light are so enormous that, if left unscreened, they will destroy visual comfort in a room. So the challenge is to employ as much of it as possible guaranteeing visual comfort.

32.1 Daylighting in DIALux

Obviously it is then a very useful feature for a lighting calculation software to evaluate daylighting.

The work flow for DIALux is:

1. Input the geographical position of the system (close to what city in the world).
1. Orient the building (clockwise from North).
2. Describe the obstructions.
3. Choose a time of the year.

The geographical position serves to obtain latitude information, which, coupled with the time of the year, gives the amount of sunlight the place will receive.

The orientation of the building describes when the windows will be illuminated, and how deep into the interior will light penetrate. Windows facing east or west will let sunlight in much deeper, as they face the sun when it is low in the sky.

The obstructions are affected by two things: what surrounds the building, and on what floor the interior is. On average a penthouse will be sunnier than a ground floor, and a detached house on top of a hill will be sunnier than an apartment in the city centre that is surrounded by other buildings.

The indicator the program calculates is called Daylight Factor.

32.2 Daylight factor

It indicates the amount of illuminance produced by natural light in the interior as a percentage of the illuminance produced by natural light outdoors.

Figure 241: Daylight properties tab.

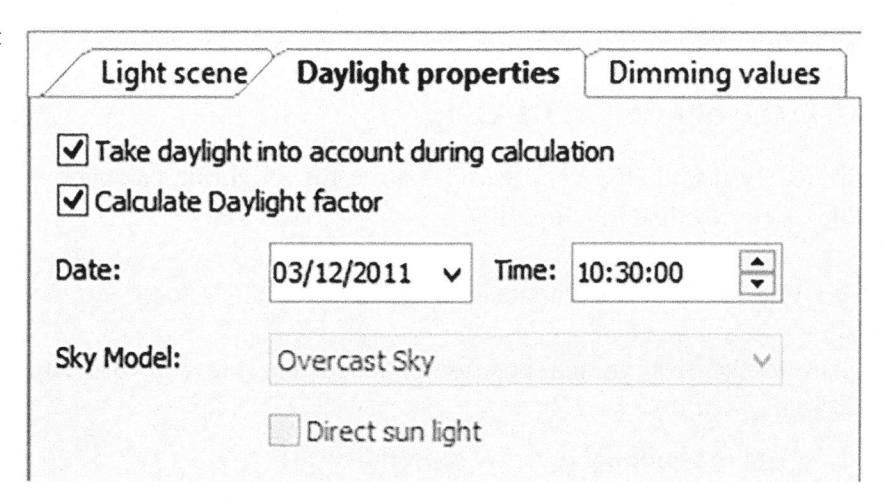

In other words it gives a quantitative measurement of how permeable the building is to natural light.

> ```
> Daylight Factor (DF) is the quotient of the
> illuminance indoors on the work plane, and
> the illuminance outdoors under a cloudy
> sky.
> ```
> (67)

A room is naturally lit if the daylight factor is larger than 0.02 (or 2%).

Daylighting

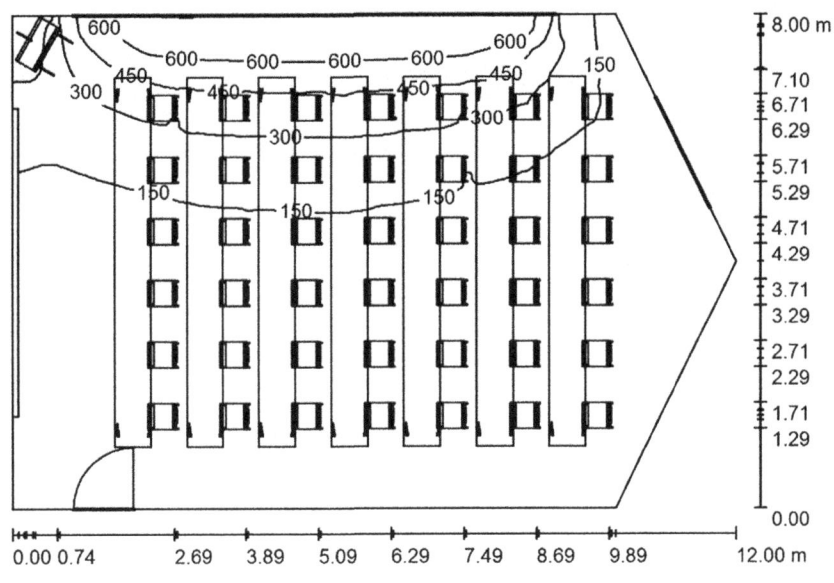

Figure 242: Summary of results for the daylight factor scene.

Surface	Reflectance.	E_{av} [lx]	E_{min} [lx]	E_{MAX} [lx]	U_0
Work plane	/	162	36	759	0.224
Floor	42	100	31	416	0.316
Ceiling	70	77	40	201	0.524
Walls	50	69	27	179	/

32.3 Calculation

We want to find out how much natural light the interior will let in, so the objective is to calculate the daylight factor. In Figure 241 we just select the date, check the **Calculate daylight factor** radio button, and start the calculation.

It is important to notice that if the radio button in the previous paragraph is checked, all the luminaire in the room will be turned off, no matter what is set in the **Dimming values** tab.

32.4 Evaluation of the results

The Summary page in the Output tree under Light scenes, Daylight with Factor contains data on which we can reflect. The two important questions to which the designer must find an answer regarding daylighting are:

1. Will natural light be able to contribute to the lighting of the interior?
2. Will natural light damage visual comfort?

Looking at Figure 242 we can answer both questions:

1. With an average for the whole interior of 162 lx and a maximum of 759 lx, daylight is definitely contributing to the lighting of the interior.
2. Because E_{MAX} is so much larger than E_{av}, we can expect that, even in December, daylight will be so powerful that it will damage visual comfort.

32.5 False colour display

In order to better answer question n.2, we can use the **False colour display** for luminance (Figure 243), which produces a simple and easy to read image of the luminance values in the surfaces of the interior.

So, from the back of the room, some students will have a luminance of over 80 cd/m² on one side of their field of view, and under 10 on the other. This is what damages visual comfort, and it is what should tell the designer to use curtains or any other way to reduce the luminance of sunlight.

Unfortunately, the false colour display for luminance has a problem: when you move the viewpoint, the values do not change. This means that the programme treats all surfaces as diffuse. This is not an issue for a plaster wall, for example, but for a metal table it is

Daylighting

unacceptable. The designer must be aware of this, and, if in doubt, use the illuminance diagram (see Figure 244 for an example) and think about the properties of each surface and how they will affect luminance.

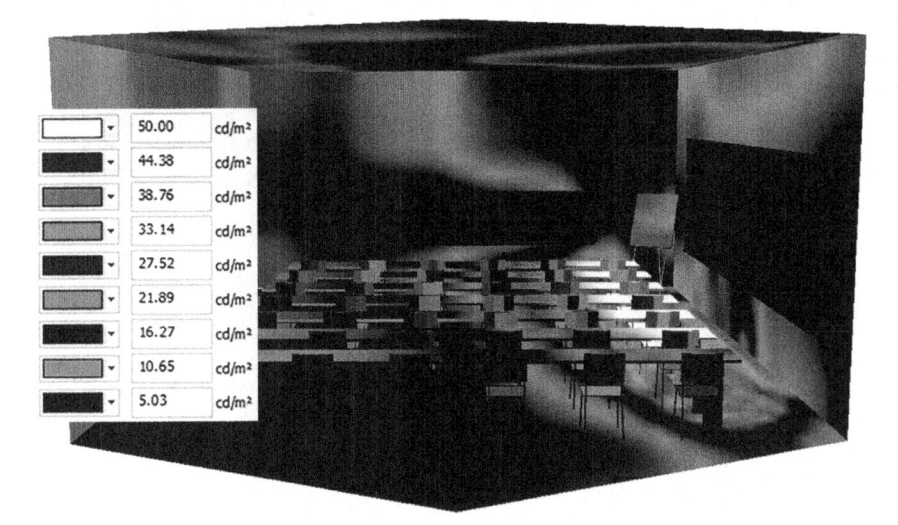

Figure 243: False colour display for luminance in the interior.

The false colour display can show illuminance as well (perpendicular to the surface). It becomes a very powerful tool to evaluate the lighting system, especially if one carefully sets the data on the false colour display legend.

I usually suggest a setting such as the one in Figure 244. I set colours to Red-Yellow Gradient, then consider the use of the interior to set the minimum value. If it is a room where the users can spend time, the minimum requirement is 100 lux (EN 12464-1, 2011), so I set the minimum illuminance to 100 lux and change its colour to black. This makes it so that any area with an illuminance of 100 lux or less will be painted black, so it is easy to find out where the system is unacceptably insufficient.

I select the maximum value at about 1.5 times the average required, or slightly less, then press **Interpolate** and have the computer calculate the points in between. I then change the colour of the maximum to white, so that all the areas that have 400 lx or more will be painted white. In those areas the illuminance is too high, and the system must be fixed.

Figure 244: Suggested settings for False Colour Display - illuminance.

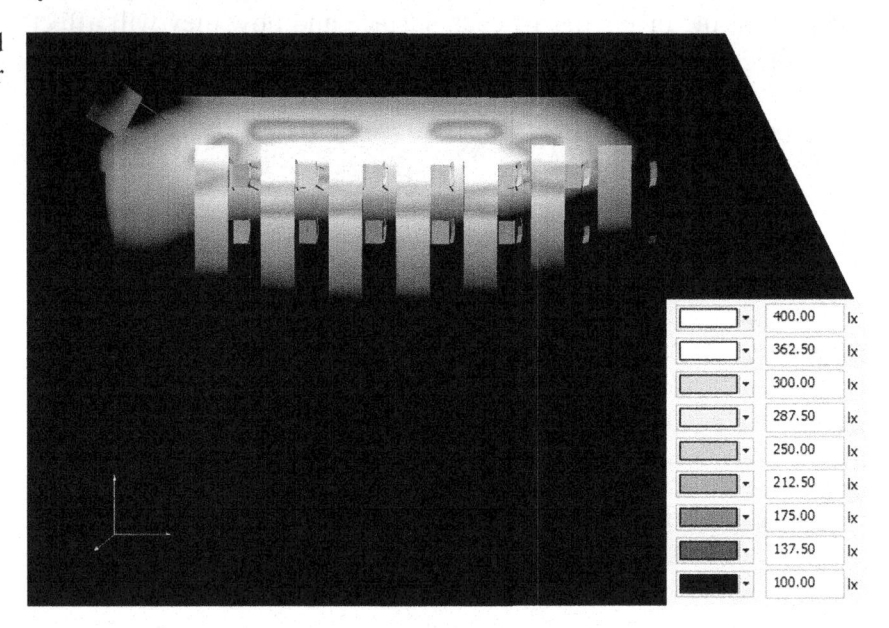

Finally I select the point closer to the average required and change it to the actual average, in this case 300 lx, then modify its colour to something very different (pink in this case), so that I can see where in the interior the illuminance reaches the average required. If most of the interior is surrounded by a pink area, I know that the illuminance will be over 300 lx; if the pink area covers most of the interior, I know I did an excellent job, because the whole of it is at or around 300 lx.

With these settings in place, I can evaluate a lighting system with just a glance.

32.6 Daylight factor isolines

I used the **Summary** of results in Figure 242, rather than the actual daylight factor isolines, because it contains more information in a single page, and because obviously isolux and daylight factor curves

Daylighting

have the exact same shape[1]. To show this, in Figure 245 there are the isolines for daylight factor.

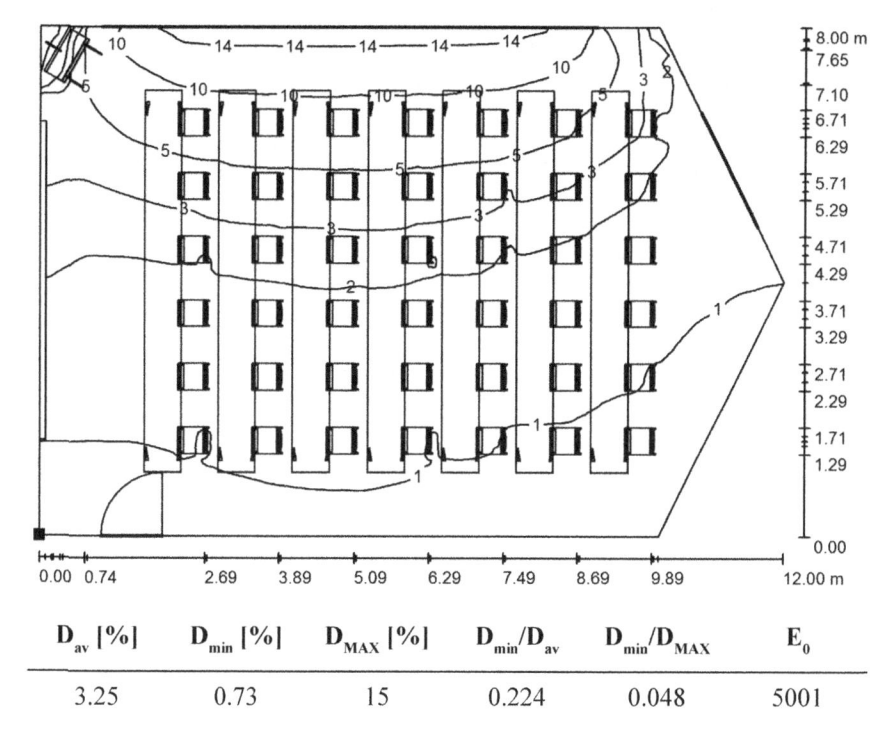

Figure 245: Daylight factor isolines.

D_{av} [%]	D_{min} [%]	D_{MAX} [%]	D_{min}/D_{av}	D_{min}/D_{MAX}	E_0
3.25	0.73	15	0.224	0.048	5001

The average DF is 3.25%, and almost half the interior has DF ≥ 2%, but the uniformity for DF is pretty bad. This means that, where it reaches, daylight can more than illuminate the interior, so the wise designer should provide some sort of screening for the windows (curtains maybe).

At the same time, daylighting is not enough to reach the whole interior, so the artificial lighting system may need to be on even during the day. This requires a very efficient light source, so that costs are kept to a manageable level.

For reference, the horizontal illuminance outdoors E_0 is 5001 lx.

---

1. Only their values are different: to get the daylight factor, one must divide the isolux value by the horizontal illuminance outdoors E_0.

Finally let's look at uniformity and average illuminance for all the surfaces we consider in this interior (Table 36).

Notice the difference between the work plane and the floor: the average is smaller because of the shading effect of all the desks and chairs; the uniformity appears better, just because the program is not calculating the total darkness that there is right under the legs of desks and chairs, so the minimum does not reach zero (it is actually 31 lx compared to 36 lx on the work plane), and it is divided by a much smaller average.

Table 36: Natural light on the different surfaces of the interior.

Surface		$\overline{E}_m$	u_0
Work plane		162	0,224
Floor		100	0,316
First row		145	0,440
Sixth Row	Task	135	0,364
	Surrounding	170	0,238
Whiteboard		93	0,677
Flip chart		140	0,755

The whiteboard is much less illuminated than the rest. The uniformity is very good, but it means nothing with such a low average.

The flip chart is the best lit surface, being close to the window and small in size.

This simulation has confirmed that this lighting system must be split into two parts: one will be turned off when natural light is present, the other will stay on. In order to do this, we must first insert luminaires into the project.

Daylighting

32.7 Daylight screening

The details of daylight control go beyond the scope of this book. It is sufficient to look into the logic behind the different devices, and two practical examples.

The objective of any device dedicated to daylight control are:

1. Improvement of visual comfort.
2. Increase of daylight factor.

Simply reducing the amount of sunlight entering the interior accomplishes objective 1. A neutral density filter film on the window, a dark glass, a curtain, or a frosted pane of glass in the window, will all reduce the natural light's luminous flux entering the room and the pool of very high illuminance (and luminance) produced by the sun close to the windows.

A curtain of a frosted glass pane will also diffuse the light more, increasing DF slightly (objective 2), but it will completely isolate the interior from the view outside. The opposite hyappens with a dark glass or a neutral density filter film: the view is uninpeded, but the penetration of natural light in the interior is not increased,

Table 37 shows what happens to the isolines when the window panes transmit only 50% of the incident luminous flux (e.g. frosted glass or curtains) compared to the 90% of clear glass.

Glass	E_{av} [lx]	E_{min} [lx]	E_{MAX} [lx]	U_0
Clear	162	36	759	0.224
Frosted or Curtain	103	15	615	0.145

Table 37: Comparison between daylighting illuminance and uniformity with a clear window and a frosted window.

Figure 247 and Figure 248 show the relevant isolines. The illuminances closer to the windows are significantly reduced, so the luminance contrast will be smaller, and with this the visual comfort will improve.

Unfortunately the penetration in the interior is reduced: the 50 lx isoline (which in this example coincides with DF=1%) is now at best in the middle of the room, and the average DF goes from 3.25% to 2.07%.

Figure 246: Sketch of a light shelf.
The sunlight is reflected towards surfaces that do not threaten the room's visual comfort while increasing DF, and the area immediately close to the window is shaded.

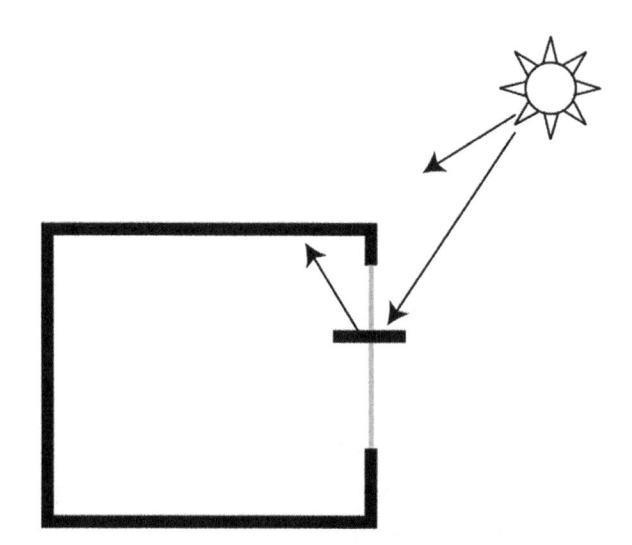

Figure 246 shows the sketch of an interesting device: a light shelf. It is a reflective horizontal element that both shades the area right next to the window and reflects sunlight to the ceiling of the interior. It must be positioned higher than eye-level to avoid glare.

32.8 Self evaluation

Answer the following questions to test your understanding of the material:

1. What is daylighting?
2. What is the daylight factor?
3. What is the false colour display feature? Describe it in detail.
4. Explain what are the objectives of daylight screening.

Daylighting

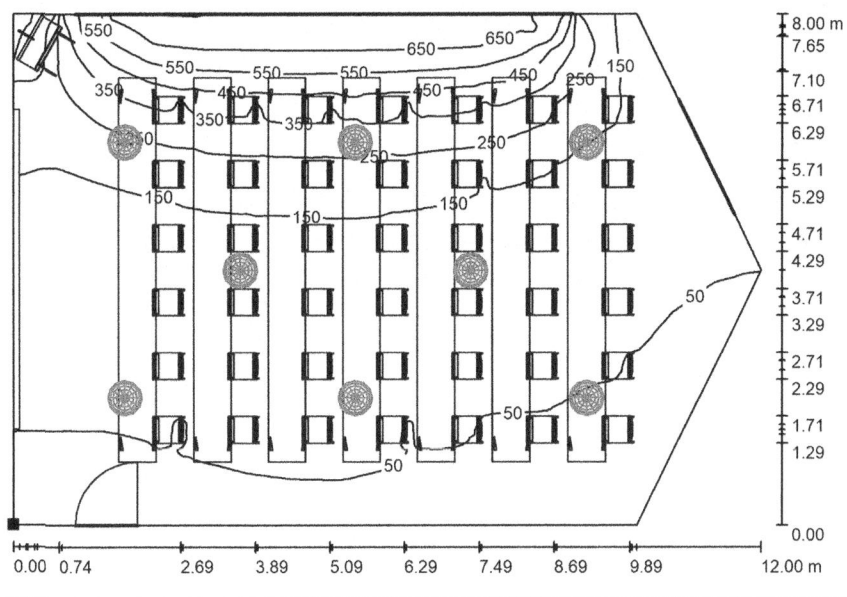

Figure 247: Isolines on the workplane with clear glass windows.

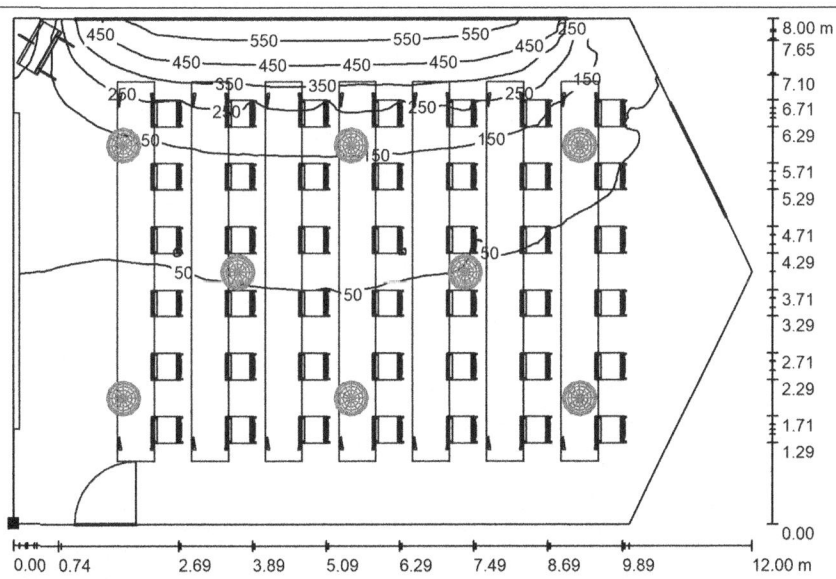

Figure 248: Isolines on the work plane with frosted glass windows

5. Describe three devices to screen an interior from direct sunlight, explaining how they accomplish the required objectives.

Figure 249: Side view of the classroom used as sample interior.

Figure 250: Plan view of the classroom used as sample interior.

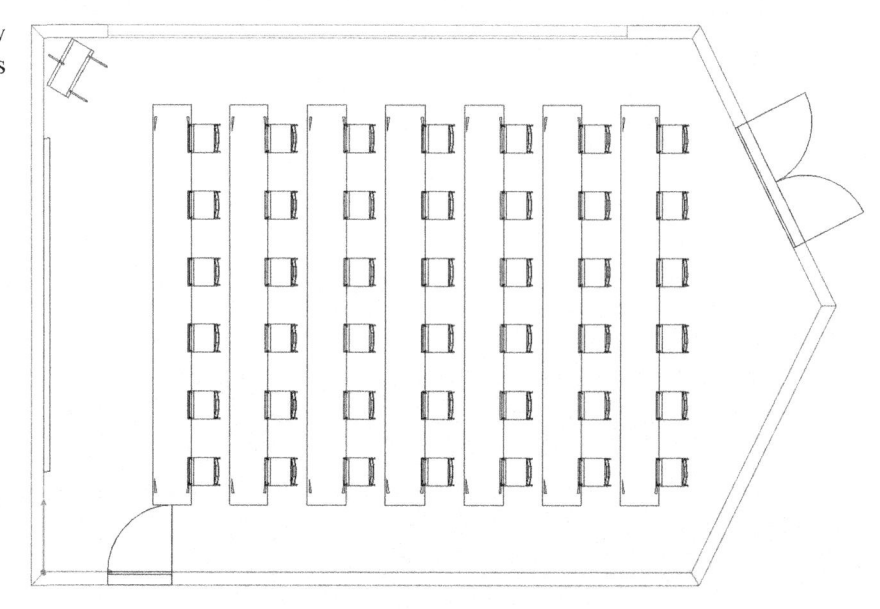

Figure 251: View of the obstructions of the classroom.

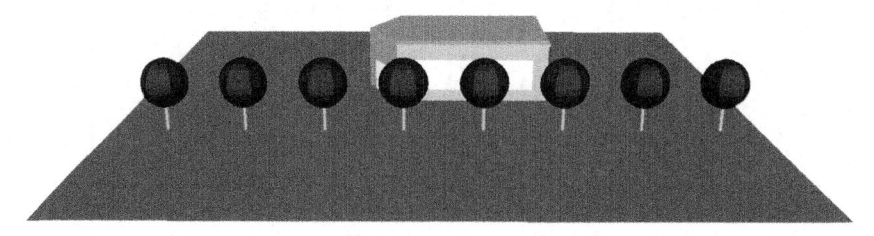

Luminaires in DIALux

33

First of all, let's review the requirements for a classroom, from the European regulations in Table 38.

$\overline{E}_m$	UGR_L	u_0	R_a
300	19	0.60	80

Table 38: Requirements for classrooms.

33.1 Insertion into our sample project

I am going to make a sub-optimal choice for this system, because it comes in handy for educational purposes.

Let's say we need to use the Artemide Nur luminaire, a 120W compact fluorescent, suspended, axially symmetric fitting, metal grey in colour, with an anti-glare screen.

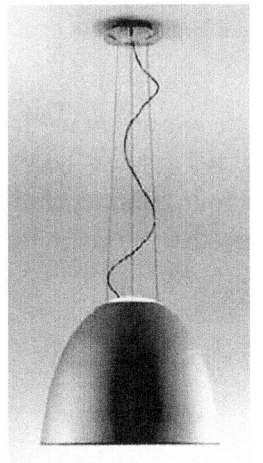

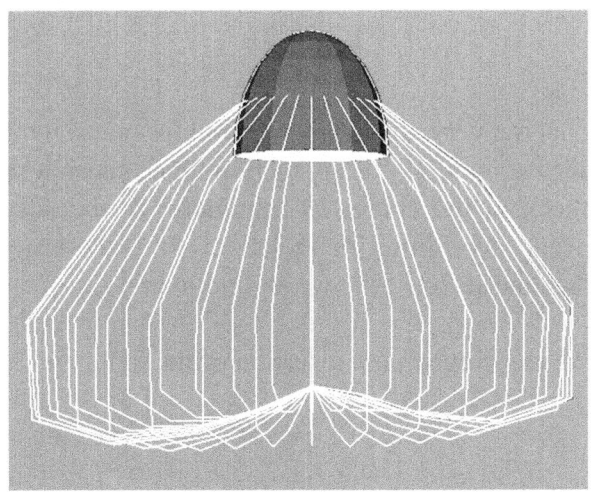

Figure 252: Picture and photometric solid of Artemide's Nur luminaire.

A good first step would be to insert a field of luminaires (`Insert`, `Luminaire Arrangement`, `Field Arrangement`), to find out how many we need to reach the required 300 lx. One just needs to type 300 in the E cell under `Rough Calculation` in the **Mounting** tab, press **Suggestion**, and see that the program proposes to use three rows of three luminaires (see Figure 253 on page 368).

Figure 253: Settings and picture of the field of luminaires in the classroom.

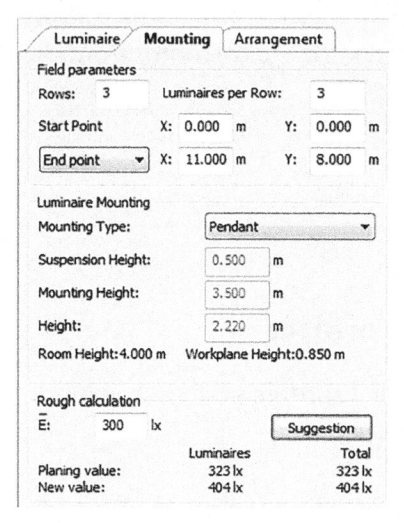

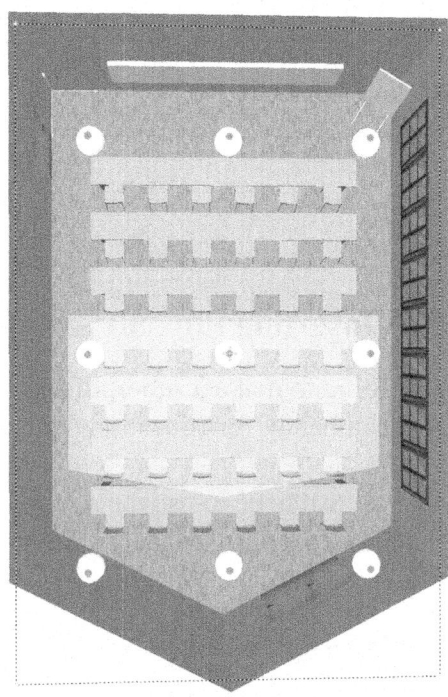

This is a very good starting point for our system: we know that we'll need around 9 fittings to illuminate the interior. It is also a good check of the choice of luminaire: if one needs four-hundred and twenty six fittings for a relatively small bookstore[1], then it is certain that one selected the wrong luminaire.

We can confirm the insertion of the field, and calculate the results.

---

1. True story...

33.2 Positioning

The calculation gives us an average of 376 lx, an uniformity U_0 of 0.160, 1.08 kW of power installed, and a Specific Connected Load (SPL[2]) of 3.26 W/m²/100 lx.

We need to examine the disposition of the luminaires and see whether this is the best possible result, or if in can be improved. The key is in the fact that this fitting is axial symmetric, this means that it will produce large round spots on the floor. Our objective must be to pack the round spots up as best as possible in a rectangular room.

The problem is equivalent to trying to pack tennis balls into a box with as many balls per box as possible (Figure 254), and the solution is the natural way in which spheres fill a box: the quincunx disposition, which, applied to the project, gives Figure 255.

Figure 254: Parallel disposition (left) and quincunx disposition (right) of tennis balls into a box.

Let's think about the problem.

---

2. For a definition of the Specific Connected Load, see 22.3 on page 262.

The old system used nine luminaires and produced 376 lx maintained on average. The new system is only using eight fittings, so, if the two dispositions were equivalent, we would expect the average illuminance to become smaller. More precisely, it should be proportional to the installed flux, so. since we are using 8/9 of the flux, or 8/9 of the fittings, we would expect the illuminance to be 334 lx, because 334 lx is the 8/9 of 376 lx.

If the new disposition is better than the old one, then the average maintained illuminance will be more than 8/9 of the old one.

Let's calculate the new system and look at the results The actual average illuminance is 357 lx, which is an improvement of more than 6%. Obviously the SPL has decreased as well, showing that the quincunx disposition is more performing than the parallel one.

Figure 255: Parallel disposition (left), and quincunx disposition (right).

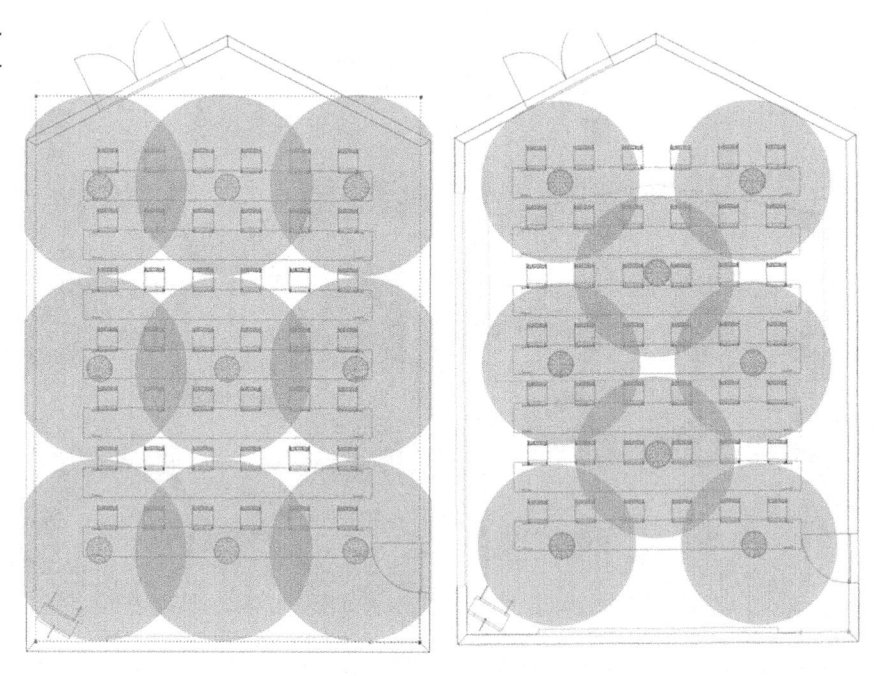

Because the system needs to be split, we will insert light fittings in lines, instead of in a field, so that we can name each line with a significant name that makes it easier to link it to a specific control group.

Indicator	Old system (parallel)	Expectation for new system	Results for new system
Average illuminance [lx]	376 lx	334 lx	357 lx
Uniformity	0.160	0.160	0.142
Power installed [kW]	1.08	0.96	0.96
Specific Connected Load [W/m²/100lx]	3.26	3.26	3.06

Table 39: The quincunx disposition is a better performer than the parallel one for this luminaire.

33.3 Definition of the necessary scenes and control groups

Both systems can be split well: the line of luminaires closer to the window will be turned off when there is natural light, and the one farther away will remain on. The fate of the middle line is still unsure, but "Figure 245: Daylight factor isolines." on page 361 makes us think that it might be off during the day as well. To confirm this, we need to calculate both cases, and choose the best one.

Lines of luminaires	Control groups	
	Always	Sometimes
Close to the window		•
Centre line		•
Far from the window	•	

Table 40: Control groups and lines of luminaires in the classroom.

The three scenes we will need to evaluate are:

1. Daylight with factor: a pure natural light scene, done in 32.3 on page 357.
2. Daylight and artificial: both natural light and part of the artificial lighting system are working to illuminate the classroom.
3. Artificial: The lighting system is fully on.

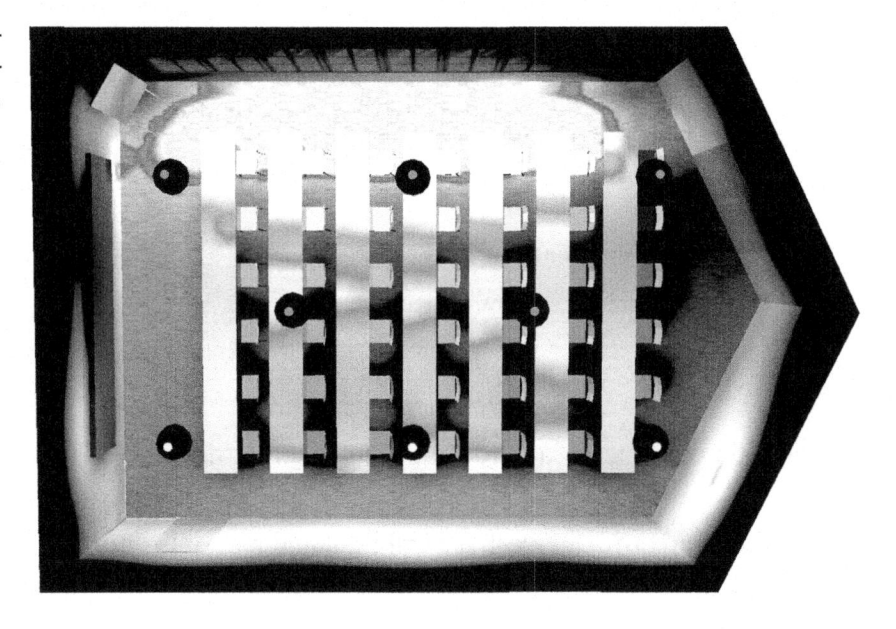

Figure 256: Isolux and illuminance values on the various tasks in the daylight and artificial scene.
The specific connected load here is 0.82 W/m²/100 lx.

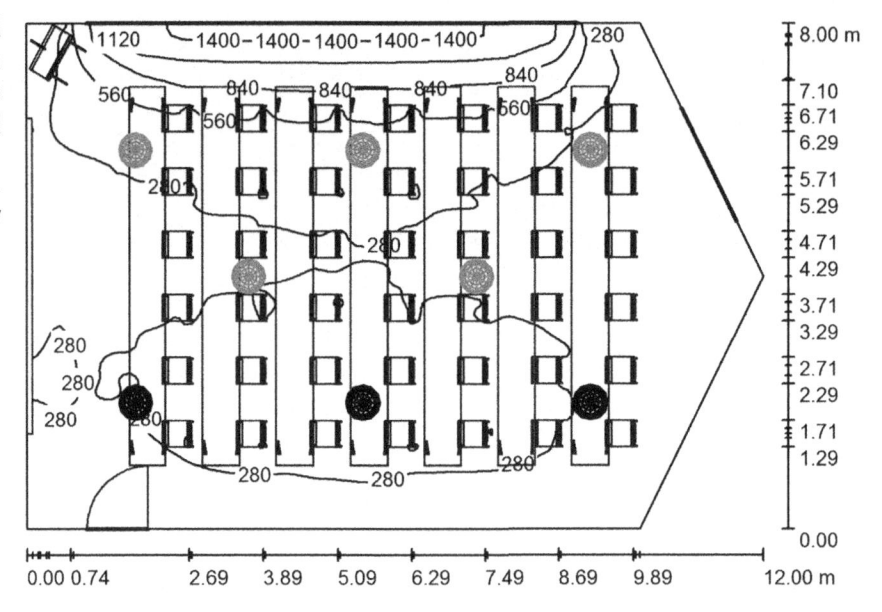

Surface	E_{av} [lx]	E_{min} [lx]	E_{MAX} [lx]	u_0
Work plane	367	50	1433	0.137
Whiteboard	239	187	298	0.782
Flip chart	274	217	351	0.790

Figure 257: False colour display in the daylight and artificial scene.

Luminaires in DIALux

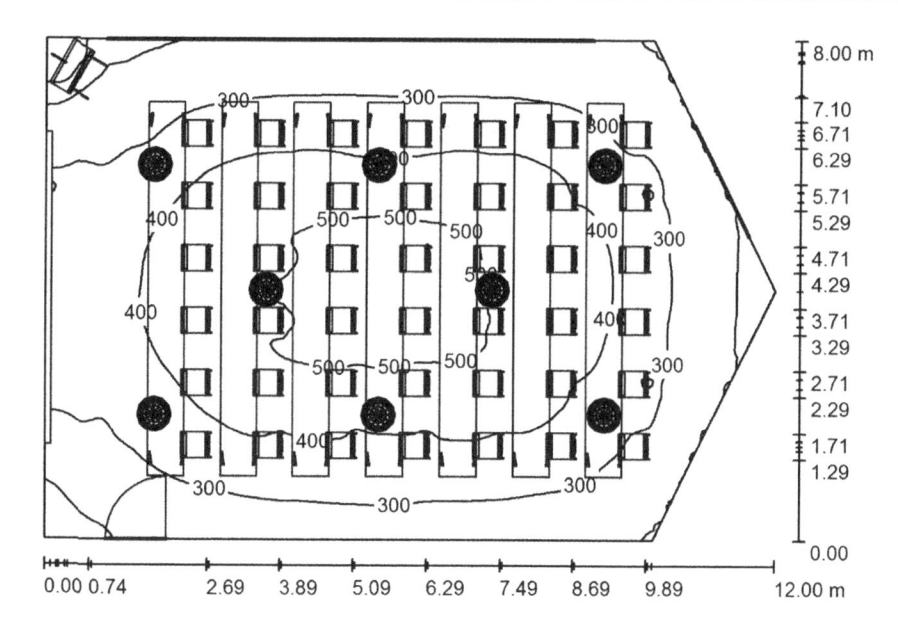

Figure 258: Isolux and illuminance values on the various tasks in the artificial scene.
The specific connected load here is 3.06 W/m²/100 lx.

Surface	E_{av} [lx]	E_{min} [lx]	E_{MAX} [lx]	u_0
Work plane	357	51	550	0.142
Whiteboard	240	203	283	0.846
Flip chart	230	202	248	0.875

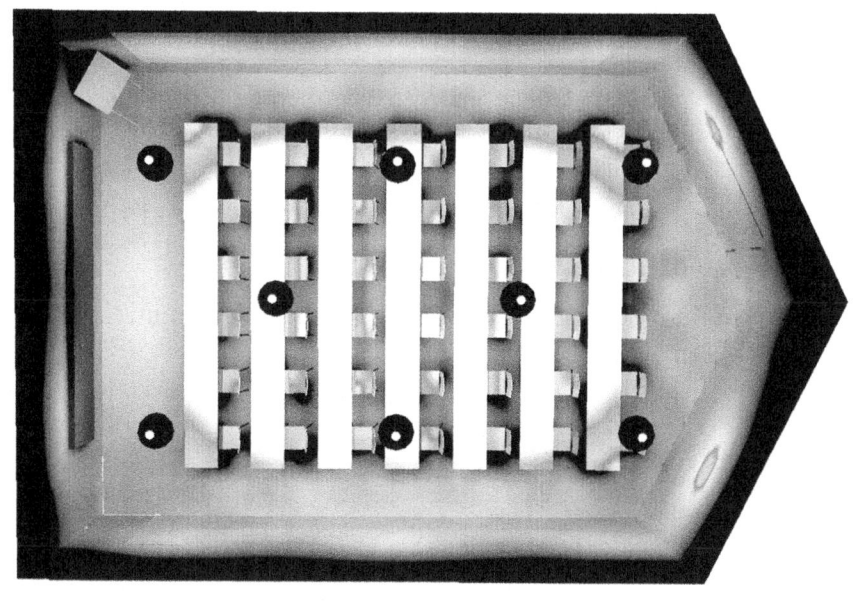

Figure 259: False colour display in the artificial scene.

Table 41: Dimming of the control groups in the scenes.

Control groups	Light scenes		
	Daylight with factor	Daylight and artificial	Artificial
Sometimes	0	0	100%
Always	0	100%	100%

I will create two control groups:

1. Always: for luminaires that need to be on even when daylight is present (as in far from the window)
2. Sometimes: luminaires that can be turned off when natural light is present.

Table 40 shows what lines of luminaires belong to what control group.

Table 41 shows the dimming of each control group in the three scenes.

33.4 Self evaluation

Answer the following questions to test your understanding of the material:

1. How can one insert luminaires into the project?
2. How can positioning affect the results.
3. How to find the best positoning when comparing multiple lighting systems.

Figure 260: Isolux and illuminance values on the various tasks of the daylight scene.
The specific connected load here is 0.82 W/m²/100 lx.

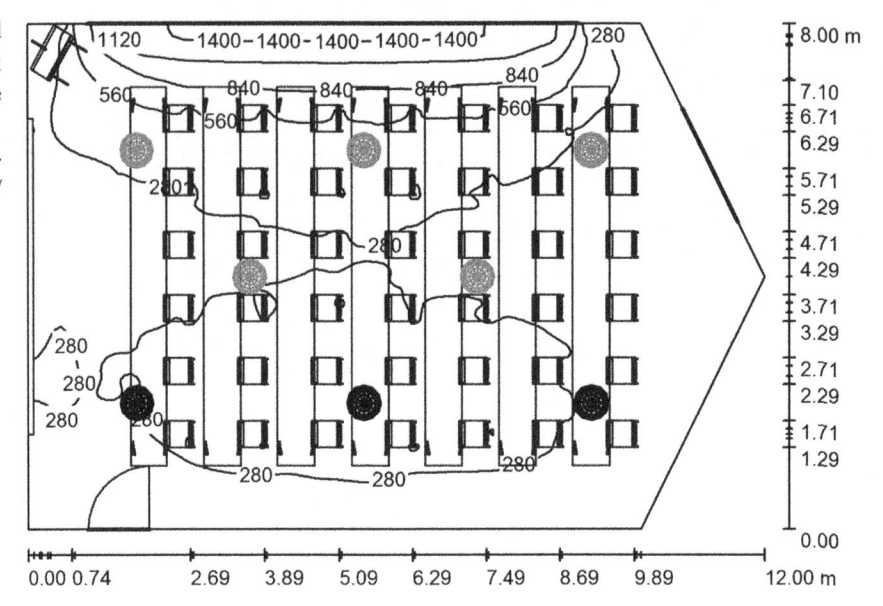

Surface	E_{av} [lx]	E_{min} [lx]	E_{MAX} [lx]	u_0
Work plane	367	50	1433	0.137
Whiteboard	239	187	298	0.782
Flip chart	274	217	351	0.790

Figure 261: False colour display of the daylight scene.

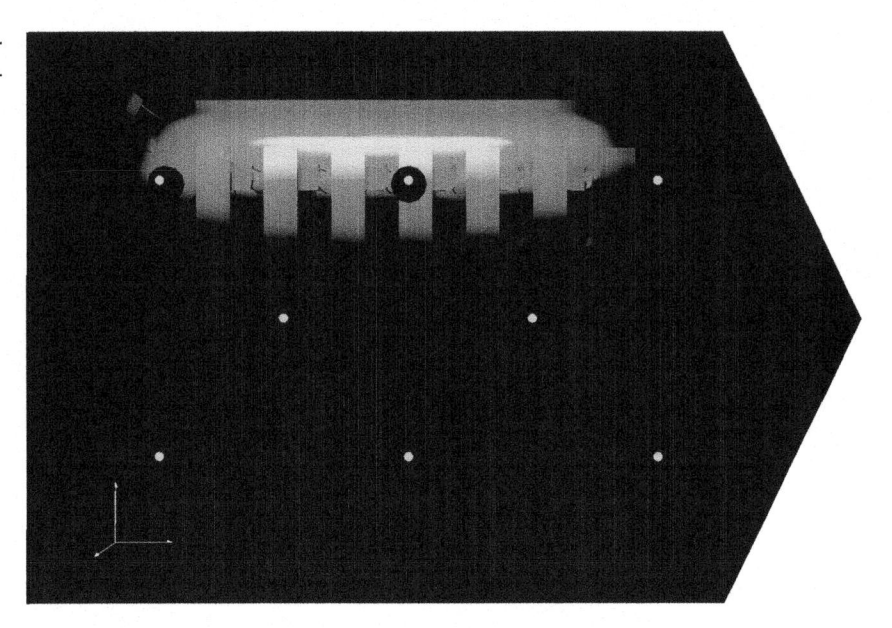

Calculation, evaluation and improvement

34

It is now time to calculate the results for all the scenes (`Output`, `Start Calculation`), then evaluate the results and propose improvements. The improvements need to be calculated and their result compared to the original to evaluate their effect.

In the next two pages there are the isolines, tables and false colour displays we arrived at in 33.3 on page 371.

34.1 Evaluation of the results

34.1.1 Daylight and artificial

The objective of the lighting system in this scene is to improve visual comfort compared to just natural light. The three lit luminaires (see Figure 260) are extending the 350 lx isolux southwards, this increases uniformity (from 0.338 to 0.545 in the `work area` calculation surface) and visual comfort.

The maximum illuminance remains very high at 1760 lx, this means that the room needs a daylight screening device (see 32.7 on page 363). The simplest way to approximate one is to change the properties of the window from clear glass to frosted glass (`Object` tree, `Window`, `Daylight properties`, `Degree of Transmission`, `Frosted Glass`), and recalculate. This reduces the transmission of the glass to 50%.

This change is enough to reduce the maximum illuminance to 1400 lx and bring the average on the work plane from 501 lx to a very acceptable 367 lx. In Figure 266 I compare the two false colour displays to show what a significant difference daylight screening makes.

Figure 262: Isolux and illuminance values on the various tasks in the daylight and artificial scene.
The specific connected load here is 0.82 W/m²/100 lx.

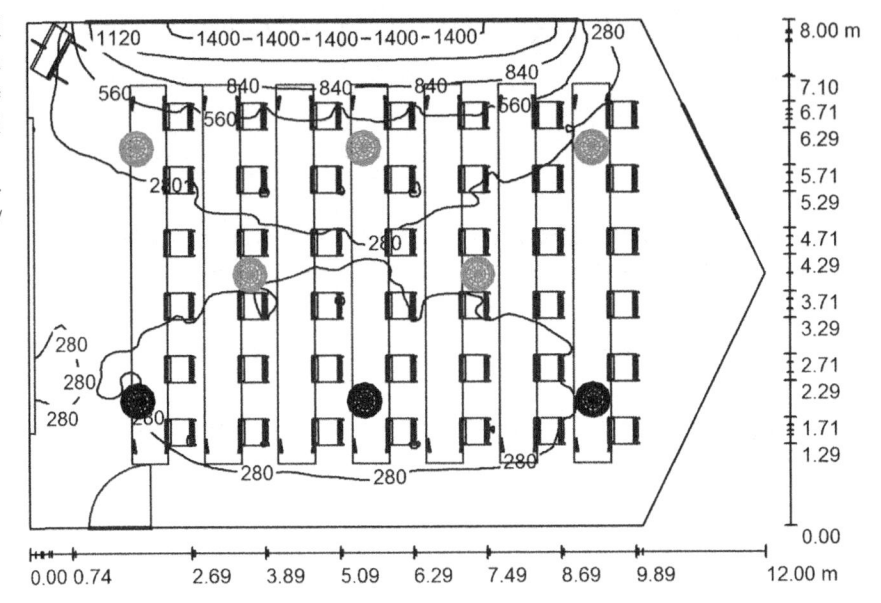

Surface	E_{av} [lx]	E_{min} [lx]	E_{MAX} [lx]	u_0
Work plane	367	50	1433	0.137
Whiteboard	239	187	298	0.782
Flip chart	274	217	351	0.790

Figure 263: False colour display in the daylight and artificial scene.

Calculation, evaluation and improvement

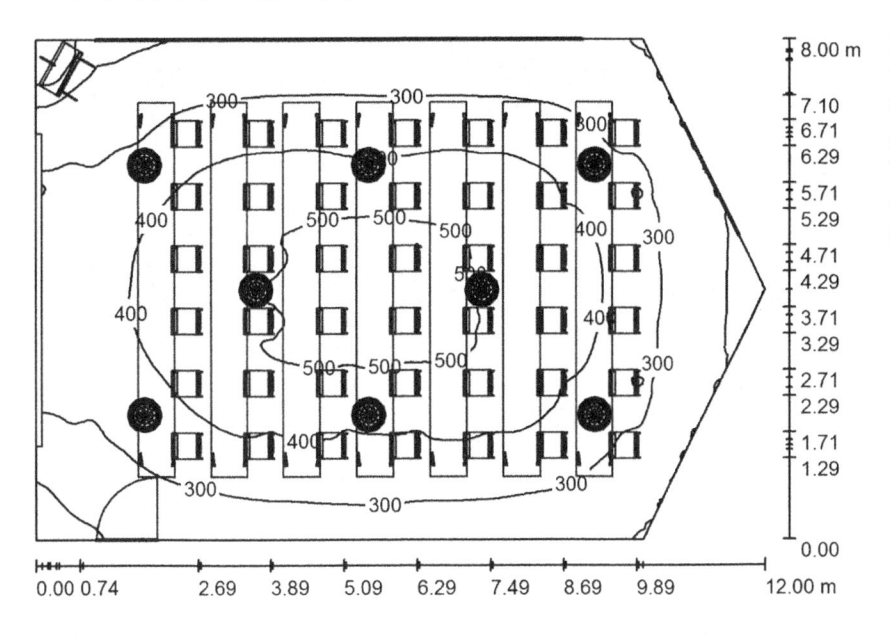

Figure 264: Isolux and illuminance values on the various tasks in the artificial scene.
The specific connected load here is 3.06 W/m²/100 lx.

Surface	E_{av} [lx]	E_{min} [lx]	E_{MAX} [lx]	u_0
Work plane	357	51	550	0.142
Whiteboard	240	203	283	0.846
Flip chart	230	202	248	0.875

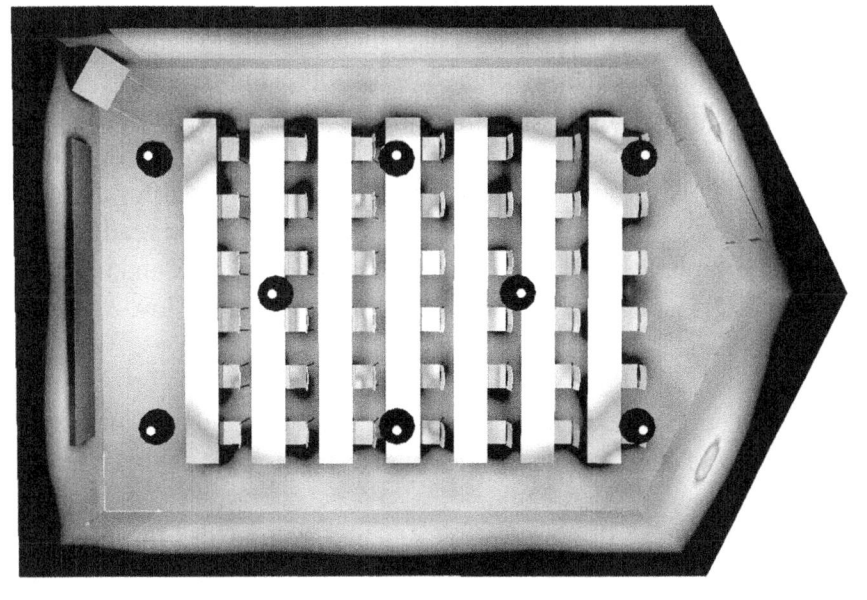

Figure 265: False colour display in the artificial scene.

380   Calculation, evaluation and improvement

Figure 266: False colour display of the illuminance in the daylight with factor scene without (top) and with (bottom) daylight screening.

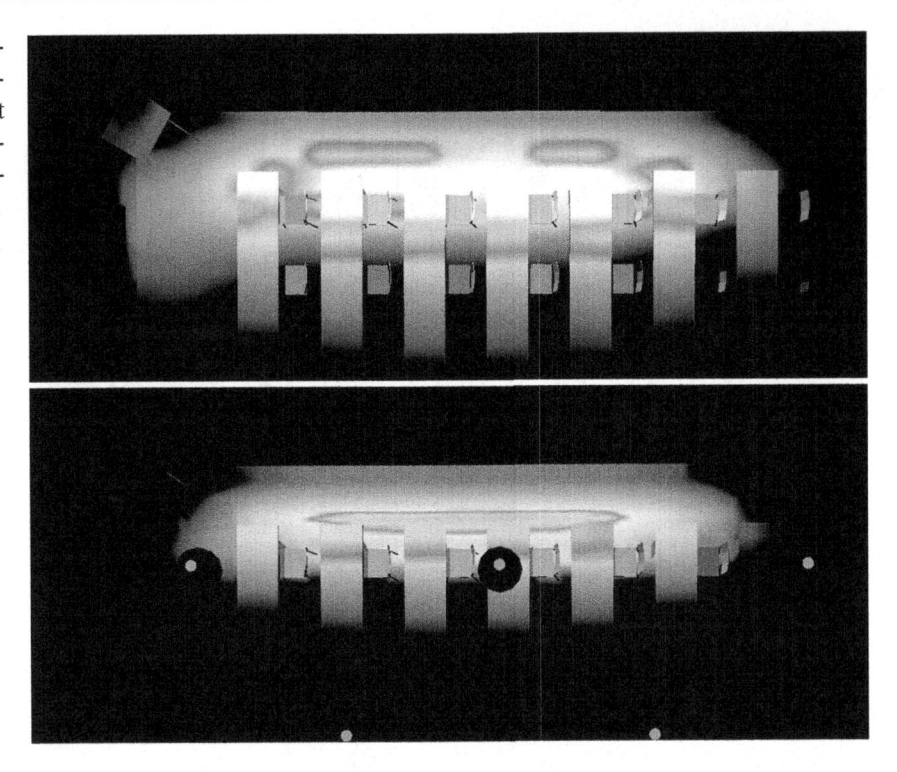

The usefulness of daylighting is well exemplified by the Specific Connected Load, at just 0.82 W/m²/100 lx, or about a quarter of what we can expect from a well designed lighting system. This excellent result comes from only having three luminaires on in an 88 m² interior in which we produce an average of 367 lx. The uniformity value is quite low, at 0.137, because of the low minimum at just 50 lx. This is rather unexpected: such a low number for an interior that should not have so dark an area at all. Before reading the next paragraph, I would suggest to the reader to go back and to study Figure 262 to find out where in the interior such a low value is reached.

If you haven't found it, look at the false colour image in Figure 263, or read the next sentence to find the answer.

The low value of illuminance is reached behind the flip chart, a moveable item that does not screen any task area in the room, but simply reduces uniformity because of the shade that is formed behind it. Removing the flip chart and recalculating puts the minimum illuminance to a more reasonable 101 lx and the uniformity on the work plane to 0.275.

Still, the calculation on the whole area only tells half the story. The horizontal illuminance over the desks is definitely important in the interior. The students will spend almost the whole time there, looking alternatively towards the whiteboard and towards their notes. The area close to the wall will only be used to enter and leave the room.

Let us then add a calculating surface that contains the desks, that is far enough from the floor that the back of the chairs do not cut through it (see Figure 267). It will be called Work Area, and it will be 8.5 m by 6 m and centred in X:5.9 m, Y:4 m and Z:0.950 m.

In Work Area, the uniformity rises to 0.482.

In Figure 263 the whiteboard does not appear to be well lit, with the average illuminance about two thirds of the work plane's. This is an area in which we need to intervene.

34.1.2 Artificial

A large part of the analysis done in the previous paragraph works here as well.

This scene (Figure 264 and Figure 265 on page 379) has a good average of 357 lx, but the uniformity is rather low at 0.142. This comes from the minimum illuminance of 57 lx that is too low.

Removing the flip chart raises the minimum to 158 lx and the uniformity to 0.441 on the work plane. Furthermore, when we consider the Working Area calculation surface, we obtain a very acceptable average maintained illuminance of 422 lx, a minimum of 265 lx and an uniformity of 0.627.

Figure 267: Work Area surface added, with illuminance values in the different scenes.

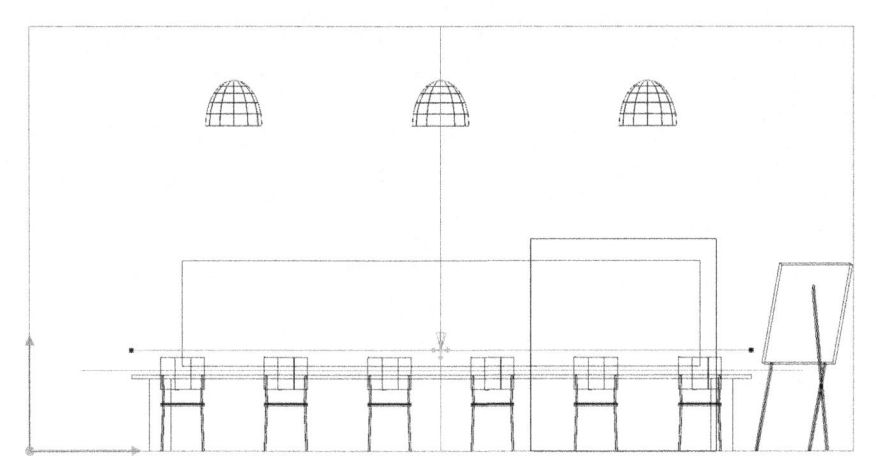

Scene	E_{av} [lx]	E_{min} [lx]	E_{MAX} [lx]	u_0
Daylight and Artificial	316	153	700	0.482
Artificial	433	269	562	0.622

It is interesting to note that even if this is not the perfect luminaire for this task, with a little manoeuvring we managed to obtain acceptable results, up to this point.

34.2 Possible improvements

Let's look in the direction of the students: the desks are rather well lit, with an average of about 430 lx, but the whiteboard has just 250 lx. Now, in order to evaluate visual comfort, we'd need to evaluate luminance, not illuminance, but unfortunately that is not yet possible in DIALux without significant approximation. Still, the difference is too high, so we need to introduce luminaires for the whiteboard.

Let's try and decide what kind of luminaire is best.

Looking at the left side of Figure 269, we see how the horizontal extension of the task is relatively large compared to the distance from the mounting point. Moreover, the need for uniformity and the fact

Calculation, evaluation and improvement

that excessive contrast between the whiteboard and the back wall is not acceptable would suggest a wall-washer.

Figure 268: Illuminance false colour display in the students' direction of view.

On the other hand, the right side of Figure 269 shows that the vertical to horizontal extension ratio of the task is small, and the mounting point position reduces the apparent surface even more, so the apparent angular size of the task from the mounting point is quite narrow.

Surface	Work area	Back wall	Whiteboard
Daylight with factor	74	34	62
Daylight with artificial	316	129	239
Artificial	433	127	240

Table 42: Comparison of illuminance between desk, back wall and whiteboard in the different light scenes.

Moreover, the Nur luminaires have a very wide beam that reaches the wall near the whiteboard. This means that we need to centre our added flux right onto the whiteboard, or we risk creating high luminance areas on the back wall (**Wall5**).

These reasons are enough to push the decision towards a row of spotlights (see Figure 270). Obviously, if the simulation were to show that the spotlight is not the right choice, the designer would go back on this decision and select a wall-washer.

Figure 269: Front view (left) and side view (right) of whiteboard.
The grey area indicates the angular size of the whiteboard as seen from the central mounting point of the luminaire.

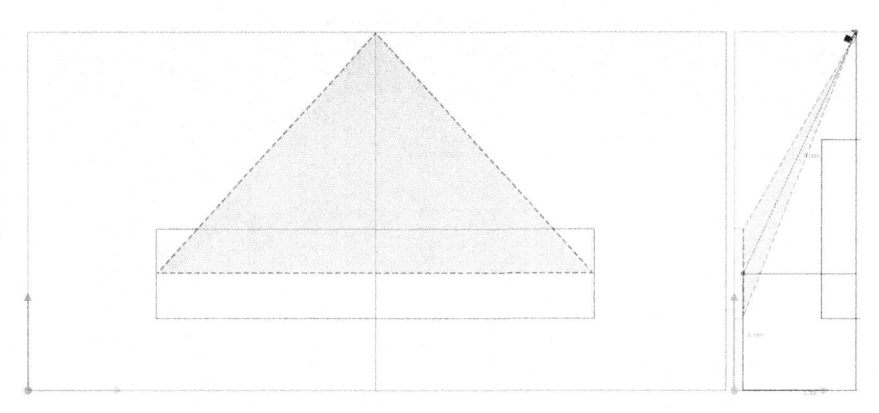

34.3 Work flow for modifying a system

Let's now look into the work flow to quickly add luminaires to a system. The best possible work flow changes with the complexity of the project, i.e. calculation time.

Figure 270: Front view of the row of spotlights illuminating the whiteboard

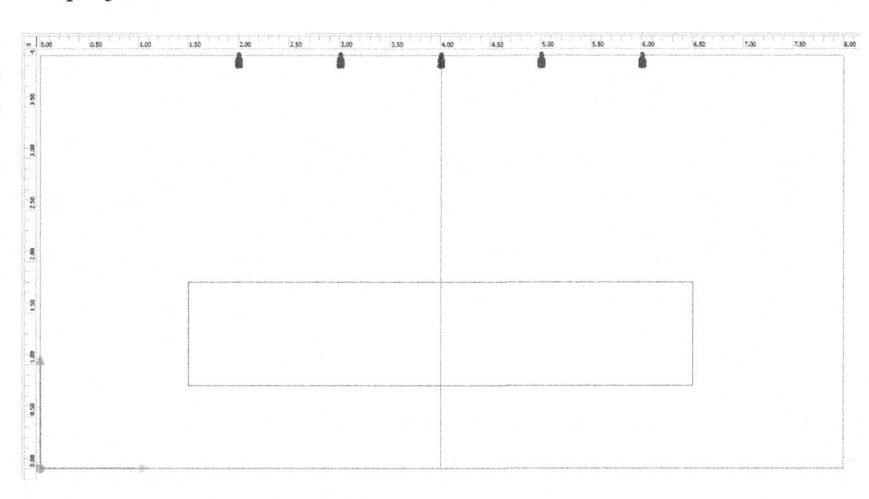

If the calculation time is short (I would consider short anything under three minutes), this is the work flow:

1. Output the results for the surface that will benefit from the improvement we are planning (in our case the whiteboard).

Calculation, evaluation and improvement

2. Duplicate the room (right click on the room, and select **Duplicate Room**), so that we do not modify the original room.
3. Apply the modification to the new room and calculate the results.
4. Compare the results of the copy to the ones in the original room.
5. If the results have improved, keep the new room.
6. Create another duplicate, and repeat the process from point 3 until you are satisfied.
7. Finally, delete the original room and rename the last duplicate.

With this method we calculate the whole room every time we make a change, and we keep the original room untouched as comparison.

If the calculation time is longer, then the best work flow changes slightly:

1. Output the results for the surface that will benefit from the improvement we are planning (in our case the whiteboard).
2. Evaluate the amount of illuminance that the modification must add.
3. While looking for the best result:
    3.1. Duplicate the room (right click on the room, and select **Duplicate Room**), so that we do not modify the original room.
    3.2. Delete all luminaires in the duplicated room.
    3.3. Place the new luminaires and calculate the results.
    3.4. Rename the duplicated room with a name that contains at least the type and number of luminaires used, the average illuminance and uniformity reached.
    3.5. Display the results of the improvement and compare them to the requirements evaluated at point 2.
    3.6. If the values have improved, keep this result.
4. Repeat from 3.

5. Once the best result is found, duplicate the original room again and add the last iteration of the upgrade to it.
6. Calculate the duplicated improved system.
7. Verify that the results are acceptable.
8. Finally delete all the intermediate steps, or, if you want to be extra careful, just rename the file.

Figure 271: Whiteboard isolux and values without added luminaires.

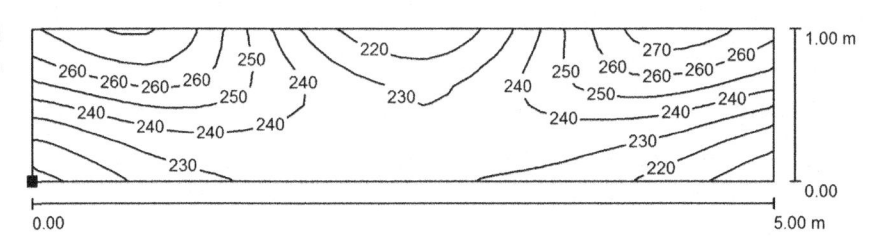

E_{av} [lx]	E_{min} [lx]	E_{MAX} [lx]	u_0
240	203	283	0,847

The advantage of this second work flow is that for every iteration the programme only calculates the smaller number of luminaires added by the improvement, rather than the whole system, with an obvious reduction in time. Only when the best improvement has been found, at point 6, the programme will calculate the whole system.

Figure 272: Whiteboard isolux and values produced by the new row of track mounted spotlights.

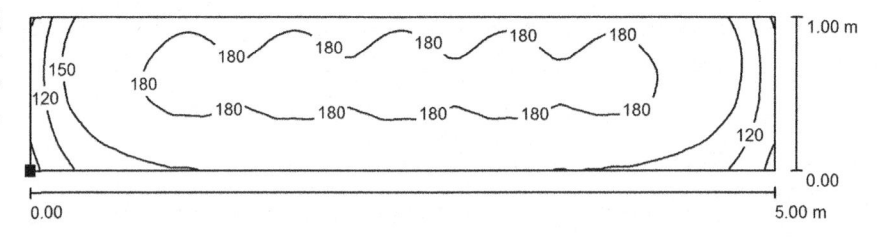

E_{av} [lx]	E_{min} [lx]	E_{MAX} [lx]	u_0
167	82	194	0,489

Let's proceed with the second type of work flow.

Calculation, evaluation and improvement 387

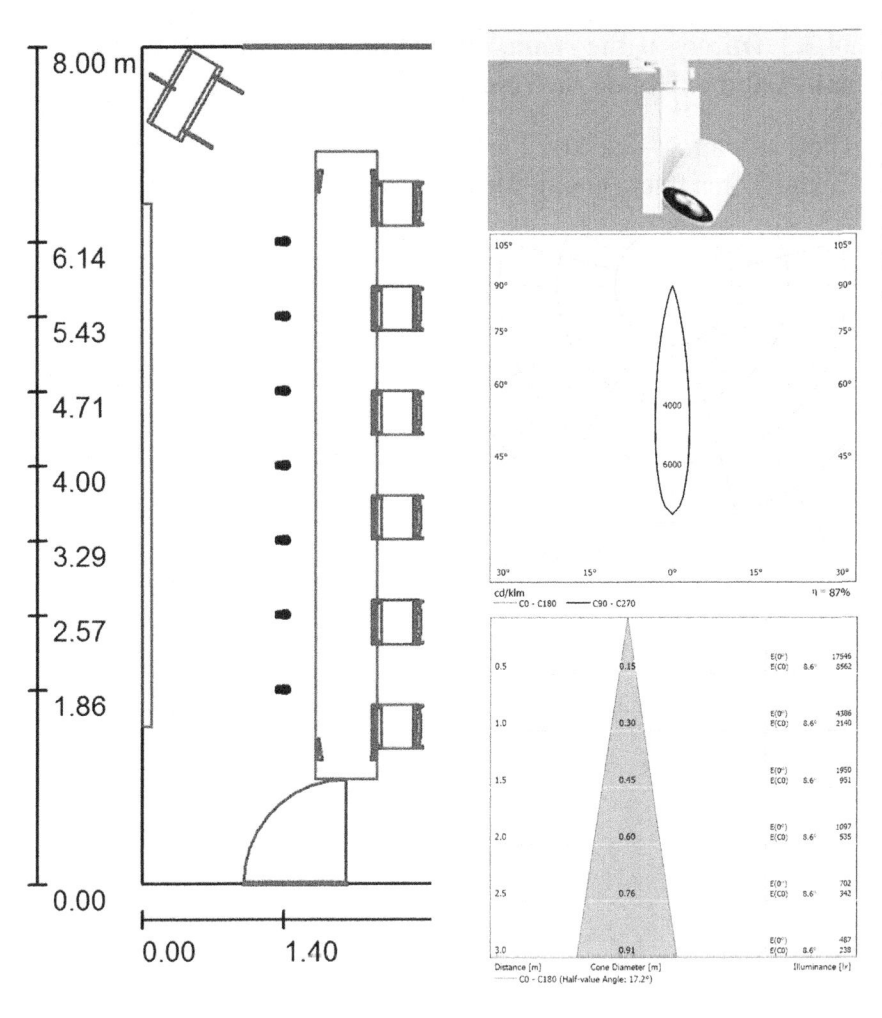

Figure 273: On the left: disposition of the whiteboard luminaires, on the left at the top: picture of the track mounted spotlight used, in the centre its light intensity distribution, on the bottom its cone diagram.

34.3.1 Output the results for the surface

The surface we need results for is **Whiteboard**, so let's calculate the project and look at the isolines and values for it (Figure 271).

34.3.2 Evaluate the amount of illuminance we need to add

In our example we want around 150 lx added. This would bring the whiteboard from 240 lx (Figure 271) to 400 lx.

34.3.3 Duplicate the room, delete luminaires, place new luminaires and calculate the result

The luminaire I selected for this part of the project is the one in Figure 273, a track mounted spotlight that is aimed towards the task.

The illuminance it provides is shown in Figure 272.

The spotlight uses a warm white 6W LED and it absorb 9W of power. It emits 570 lm of flux, with an efficacy of 63 lm/W. It is mounted at 140 cm from the wall, and the seven luminaires are evenly spaced in front of it. It is aimed towards the whiteboard with an angle of 24° from the vertical.

Figure 274: Whiteboard isolux and values produced by the updated lighting system.

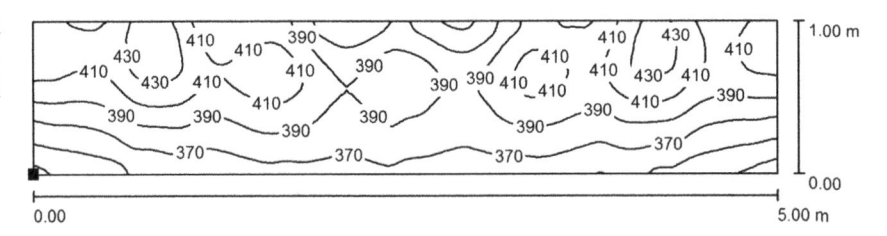

E_{av} [lx]	E_{min} [lx]	E_{MAX} [lx]	u_0
390	323	444	0,826

34.3.4 Final result

The final result for the whiteboard is in Figure 274. The addition of the row of narrow spotlights has increased the average to 390 lx, with very good uniformity ($u_0 = 0,826$) for a surface of 5 m² with an added power of nine watt per luminaire, so a total of 63 W. We calculate the Specific Connected Load of the additional system (see (68)) at 7.5 W / 100 lx /m², which is not bad at all if we consider that it is obtained using a narrow beam spotlight. The values for general lighting can be about half as large as this result, but task lighting has higher requirements in terms of luminous intensity control, which in turn means a reduced efficacy of the luminaire.

Calculation, evaluation and improvement

This confirms that the SCL can work as an indicator of the fit between the luminaire and the task. Given the same light source, so that the different efficiencies do not come into play, a fitting with a lower SCL will illuminate the task better.

$$SCL = \frac{\text{Power installed [W]}}{\text{Illuminance [100 lx]} \times \text{Surface [m}^2\text{]}} \tag{68}$$

$$SCL = \frac{63 \text{ [W]}}{1.67 \text{ [100 lx]} \times 5 \text{ [m}^2\text{]}} = 7.5 \frac{W}{100 \text{ lx } m^2}$$

The last indicator of a good system is the UGR, or the visual comfort of the interior.

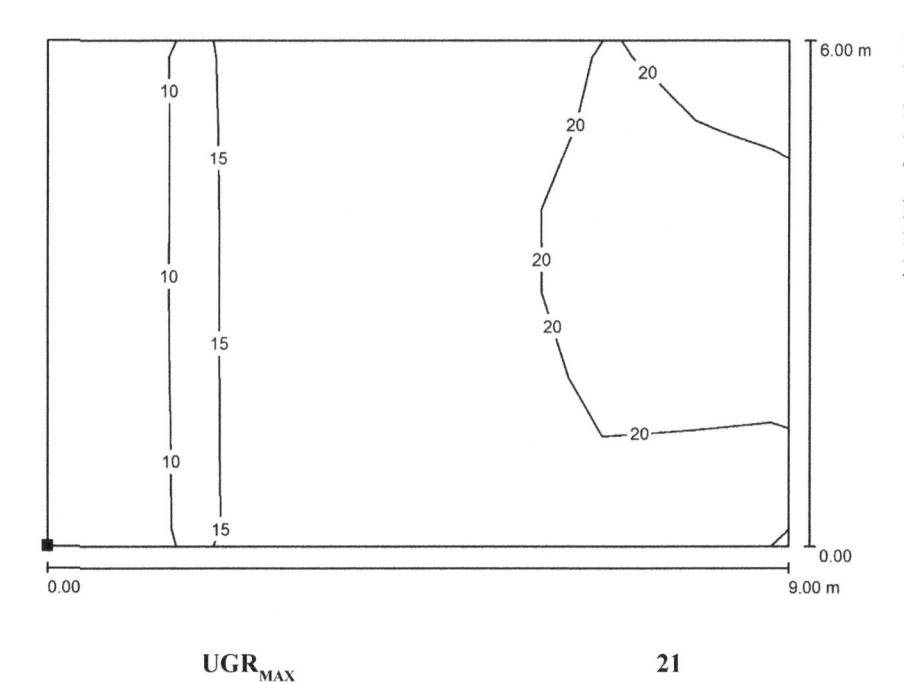

Figure 275: Isolines for UGR values in the classroom, and maximum value.
The system is not compliant to the regulations in Table 38 on page 367.

UGR_{MAX} **21**

34.4 UGR evaluation

Normally the designer should evaluate the UGR table as soon as the first calculation is completed, because compliance to UGR requirements means that a good luminaire for the interior has been chosen.

We look at the UGR table very late in the work flow, not because this is the correct time to do it, but because I feel that looking at it earlier might have hindered my explanation.

We managed to reach compliance (see Table 38 on page 367) on average maintained illuminance, uniformity and colour rendering index. It is now time to check visual comfort.

We place a **UGR Calculation Area** surface in the room, over the desks (**Object** tree, **Calculation Surfaces**). Its origin will be at 5.9 m, 4 m, 1.2 m (for a sitting observer; for a standing one change it to 1.6 m), its length 9 m and its width 6 m. Finally, we need to modify the **Observer's Viewing Direction** (**Observer** tab) to 180°, or towards the whiteboard, because that is the main direction in the room.

Figure 276: A better luminaire for the classroom.

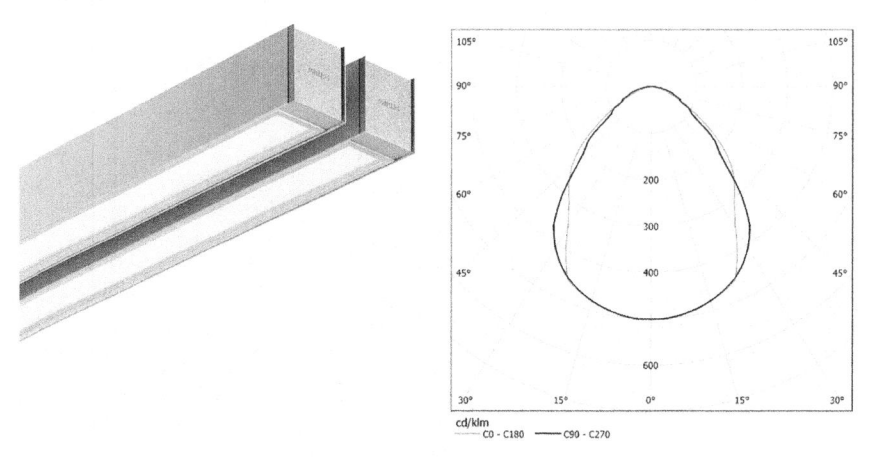

The maximum value for the UGR in the room is 21, larger than 19 (the maximum acceptable by the requirements), so the system is non compliant.

We must intervene in the interior, to reach compliance. Because this is a room that does not contain items that easily damage visual comfort, such as mirrors or other glossy reflective vertical surfaces or some such, we must modify the lighting system. As I stated at the beginning, the selection of this type of luminaire for the interior is

Calculation, evaluation and improvement 391

not optimal, and we have reached a point where the limits of the luminaire lose us the compliance.

UGR was defined in 9.4.4 on page 114. The important part of that definition is that, once the viewer's position has been defined, the luminance contrast of both luminaires and surfaces of the interior increase the UGR value.

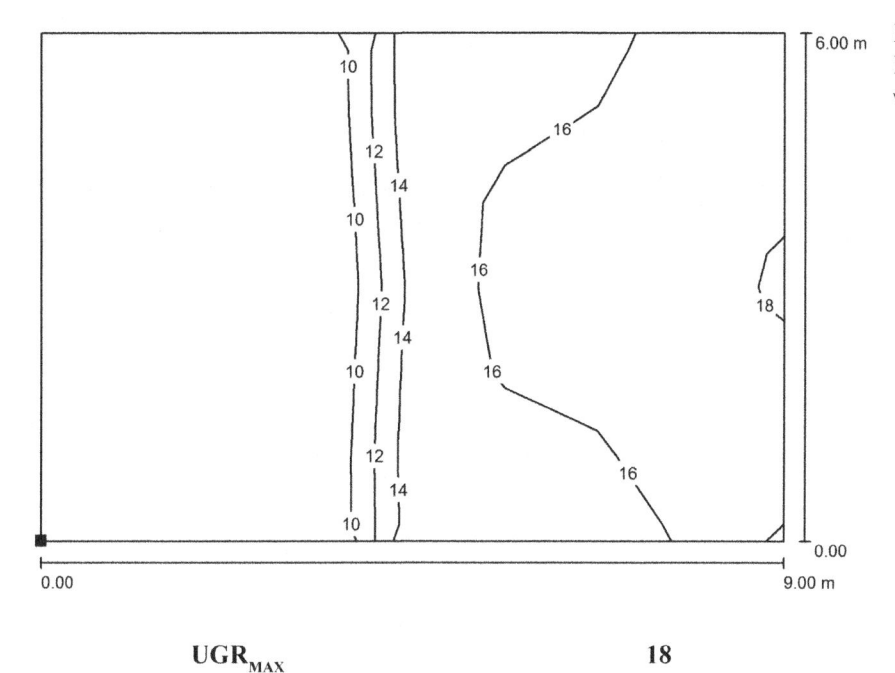

Figure 277: UGR isolines for the classroom with the new luminaire.

UGR_{MAX}	18

So, in order to reduce the maximum UGR for the interior, we need to reduce luminance contrast.

Since we can not easily look at the real luminances on the surfaces of the interior, I suggest we look at the false colour - illuminance diagram, and remind ourselves of the finishing for each surface, so that we can find out the amount of luminance generated on them.

Regarding the luminaire, we shall select one with lower luminance, and a slightly narrower beam, to reduce the unwanted luminances on the walls. A better luminaire for this interior is shown in Figure 276.

With this luminaire, our results change slightly, and we reach total compliance.

34.5 Self evaluation

Answer the following questions to test your understanding of the material:

1. Describe the analysis of the results of a calculation.
2. Explain the two different work flows for applying modifications to a system efficiently.
3. Describe two different projects and explain which method of modification works better in each.
4. Describe how to perform an UGR evaluation in DIALux.

Calculation, evaluation and improvement

Alphabetical list of topics

A

activity table 305
ampere (A) 33
angle
  plane 47
  solid 48
apparent surface 98
architectural features 299
asymmetric 241
  traditional asymmetric 244
  wall washers 248
    wide beam 252
average observer 14
average rated life 129
axially symmetric luminaires 213
  narrow beam 215
  wide beam 226

B

background area 193
ballast 131
  electromagnetic 131
  electronic 132
Beam spread 57
bright lamps 111

C

candela (cd) 49
candle over square metre (cd/m2) 99
capacitor 131
C, gamma 51
CIE classification 197
  direct 198
  direct-indirect 208
  general-diffuse 206
  indirect 200
  sample room 197

semi-direct 203
semi-indirect 204
colour
  colour bleeding 301
  extremely high colour fidelity 27
  Perception colours 13
  Spectral colours 13
  synthesis
    additive 17
colour rendering 22
commercial classification 271
  accent fixtures 278
  commercial fluorescent 273
    fluorescent strip - batten 274
    wraparound diffuser 273
  cove lights 279
  decorative lights 280
  downlight 271
  industrial fixtures 275
  linear systems 276
  task lights 280
  troffer 272
  wall grazer 277
  wall-washers (commercial definition) 276
cone diagram 84
correlated colour temperature
  25
  cool
    25
  Kelvin
    24
  neutral
    25
  reciprocal mega Kelvin
    26
  warm
    25
Cuttle model
  attributes 287
    brightness 287
    clearness 288
    flicker 288
    gloss 288
    hue 288
    pattern 288

saturation 287
texture 288
modes of appearance
  located
    illuminant 286
    illuminated 287
    object mode 287
  non located 287
    illuminant 287
    illuminated 287

D

daylight factor 356
  isolines 360
DIALux
  calculation surfaces
    free surfaces 343
    surface of an object 342
    task area and surrounding area 343
    walls, floor and ceiling 341
    work plane 341
  control group 348
  daylight factor. See daylight factor
  evaluation of the results 377
  false colour display 358
    illuminance 359
    luminance 358
  light scene 348
  luminaire arrangement 367
    field arrangement 367
  parts
    3D modeller 318
    database 321
    output package 322
    simulator 322
  possible improvements 382
  UGR evaluation 389
  wizards
    example 329
  work flow for modifying a system 384
dichroic reflector 137
diffuser 42, 189

E

electrical current 33
electric arc 124

F

flicker 108
fluorescence 123
fluorescent lamp 145
  compact integrated 150
  compact non-integrated 153
  Inner workings 146
  Lamp types 145
  traditional fluorescent 149

G

glare 109
  disability glare 109
  discomfort glare 110

H

halogen lamp 135
  without reflector 136
  with reflector 136
  halogen cycle 138

I

illuminance 62
  cylindrical illuminance 64
    average maintained cylindrical illuminance 67
    with narrow beam spots 224
  horizontal illuminance 64
  illuminance and intensity: general formula 82
  Illuminance under a luminaire 79
  illuminance uniformity and light intensity distribution 86
  illuminance uniformity U0 68
  semi-cylindrical illuminance 65
    with narrow beam spots 223
  spherical illuminance
    with narrow beam spots 225
  vertical 219
  vertical illuminance 64
incandescence 123
infrared radiation 7

IP rating 189
isolux curves 73

K

Kruithof diagram 28

L

lamps
  bulb 126
  cap 126
  cost comparison to the halogen
    compact fluorescent 150
    LED 170
    metal halide 159
  evaluating grid
    fluorescent 147
    halogen 139
    LED 177
    metal halide 162
layers
  decorative 309
  general layer 312
  task layer 310
LED
  average life 175
  binning 175
  coloured light modules 172
  driver 174
  heat dissipation 174
  inner workings 173
  lamps 169
  luminance control 177
  white light modules 172
light 6
Light intensity distribution curve 53
light output ratio 42
lumen (lm) 38
luminaire files 322
luminance 99
  luminance contrast 101
luminous efficacy 39, 44
luminous flux 38
luminous intensity 49
lux (lx) 62

M

maintenance factor 66
metal halides 157
modelling 71
mounting
  ceiling 266
  portable 269
  recessed 270
  suspended 268
  track 270
  wall 267

O

observer 14
offending zone 115
orientation of a luminaire 238

P

photometric curve. See Light intensity distribution curve
photometric solid 50
photopic spectral luminous efficiency function 14
power 34

R

radians (rad) 47
rated lumen maintenance life 129
reflectance 91
reflection
  diffuse 93
  mixed 94
  specular 92
reflector 42, 188
refractor 42, 189

S

semiconductor physics 124
shade 107
Specific Connected Load 262
spectral power distribution 9
starter 131
steradians (sr) 48
surrounding area 193

symmetric about two planes 231, 233
  narrow beam 233
  wide beam 234
    luminance control 240

T

task area 62
  identification of 306
total efficacy 43
transformer 131

U

ultraviolet radiation 8
unified glare rating 116

V

veil 105
vision
  photopic 16
  scotopic 16
visual comfort 105
visual task 61
Volt 33
voltage 33

W

Watt 34

Bibliography

Brown, B. (2008). Motion picture and video lighting. Focal press.

CIE. (1989). 81:1989 Mesopic photometry.

CIE. (2005). 97:2005 Guide on the maintenance of indoor electric lighting systems.

CIE. (1924). Proceedings of the Commission Internationale de l'Eclairage. Geneva.

CIE. (1951). Proceedings of the Commission Internationale de l'Eclairage. Paris.

Cuttle, C. (2008). Lighting by design. Architectural Press.

Cuttle, C. Modes of appearance and perceived attributes in architectural lighting design. 24th CIE session, (pp. 116-118). Warsaw.

Elmer, W. B. (1980). Optical design. John Wiley and sons.

EN 12464-1. (2011). Lighting of work places, Part 1: Indoor work places.

EN 12464-2. (2014). Light and lighting. Lighting of work places. Outdoor work places.

EN 12665. (2011). Basic terms and criteria for specifying lighting requirements.

EN 13032-1. (2004). Measurement and presentation of photometric data of lamps and luminaires.

EN 60529. (2001). Degrees of protection provided by enclosures.

IEC 60364. (n.d.). Electric installations for buildings.

IESNA LM-80. (2008). Approved method for measuring lumen depreciation of LED light sources.

Jiao, J. (2011, October). Understanding the difference between LED rated life and lumen-maintenance life. LEDs magazine, 51.

Karandikar, R. V. (1955). Luminance of the Sun. Journal of the Optical Society of America, 45 (6), 483-488.

Kruithof, A. A. (1941). Tubular luminescence lamps for general illumination. Philips Technical Review, 6, 65-96.

McCandless, S. (1932). A method of lighting the stage. Theatre art books.

Schreuder, D. (2008). Outdoor lighting: physics, vision and perception. Springer.

Smith, G. (2002, April). Disability glare and its clinical significance. Optometry Today.

United States Environmental Protection Agency. (2010). Health effect of overexposure to the sun.

Bibliography

Made in the USA
Middletown, DE
05 November 2019